Capillary Gas
Chromatography in
Food Control and Research

Capillary Gas Chromatography in Food Control and Research

edited by
R. Wittkowski and R. Matissek

TECHNOMIC
PUBLISHING CO., INC.
LANCASTER · BASEL

Capillary Gas Chromatography in Food Control and Research
a **TECHNOMIC** publication

Published in the Western Hemisphere by
Technomic Publishing Company, Inc.
851 New Holland Avenue, Box 3535
Lancaster, Pennsylvania 17604 U.S.A.

Distributed in the Rest of the World by
Technomic Publishing AG
Missionsstrasse 44
CH-4055 Basel, Switzerland

Printed in the United States of America
10 9 8 7 6 5 4 3 2 1

Main entry under title:
  Capillary Gas Chromatography in Food Control and Research

A Technomic Publishing Company book
Bibliography: p.

Library of Congress Catalog Card No. 92-62247
ISBN No. 1-56676-006-2

Copyright © 1990 by B. Behr's Verlag GmbH & Co.
Averhoffstrasse 10, D-2000 Hamburg 76
Federal Republic of Germany. Licensed edition by
arrangement with B. Behr's Verlag GmbH & Co.

Title of the first edition:
  Capillary Gas Chromatography in Food Control and Research
Copyright © 1990 by B. Behr's Verlag GmbH & Co.
Averhoffstrasse 10, D-2000 Hamburg 76
Federal Republic of Germany

Preface

Owing to the extremely complex and varied composition of foods, food analysts require highly effective separation systems. No wonder that capillary gas chromatography has easily been accepted in this field. Today's instrumental variety with its individual system units permits an almost optimal adjustment of instrument parameters to the problem to be solved. This is especially true for classical GC-analyses with regard to quick and safe results on type and quantity of value-determining food constituents as well as for monitoring legal maximum values for permitted additives in food control. Moreover, this method proved to be effective in situations involving challenges based on new problems and is now successfully applied in the receiving and process control for raw material and foods as well as in the vast field of trace and ultratrace analysis of environmental contaminants.

The intention of publishing the present book in spite of the fact that many books on gas chromatography are already available was to offer a practice-oriented work for food analysts. It offers a comprehensive survey on the possibilities, applications, and new developments of capillary gas chromatography for the complete range of examinations of food and raw material. Special emphasis was given to a generally comprehensible theoretical treatise of the subject and to restrict it to a spectrum that is essential for the daily practice of a chromatographer.

This book is written particularly for food scientists/chemists, food technologists, nutritionists, biotechnologists, analytical chemists, and veterinarians — generally all persons whose field of activity is closely connected with matrix-oriented analysis.

We would like to thank the authors who have contributed to the book. We appreciate the care taken and the time spent over what it is hoped the reader will find to be a varied and instructive compilation of chapters. We also are grateful to the publisher for giving us the opportunity to edit this book.

Berlin/Cologne

R. Wittkowski
R. Matissek

Dedicated to Prof. Dr. Werner Baltes

Authors and Editors

Editors

Dr. Reiner Wittkowski
Max von Pettenkofer-Institut
Bundesgesundheitsamt
Postfach 33 00 13
D-1000 Berlin 33

Prof. Dr. Reinhard Matissek
Lebensmittelchemisches Institut
des Bundesverbandes der
Deutschen Süßwarenindustrie e. V.
Adamsstraße 52–54
D-5000 Köln 80

Dr. Gary Takeoka
United States Department
of Agriculture
Western Regional Research Center
800 Buchanan Street
Albany, CA 94710, USA

Dr. Stefan Vieths
Institut für Lebensmittelchemie
Technische Universität Berlin
Gustav-Meyer-Allee 25
D-1000 Berlin 65

Authors

Dr. Karl-Heinz Engel
Max von Pettenkofer-Institut
Bundesgesundheitsamt
Postfach 33 00 13
D-1000 Berlin 33

Dr. Otmar Fröhlich
Lehrstuhl für Lebensmittelchemie
Universität Würzburg
Am Hubland
D-8700 Würzburg

Dr. Christiane Fürst/Dr. Peter Fürst
Chemisches
Landesuntersuchungsamt
Nordrhein-Westfalen
Sperlichstraße 19
D-4400 Münster

Dr. Rolf Hardt
Versuchs- und Lehranstalt
der Brauerei in Berlin
Seestraße 13
D-1000 Berlin 65

Table of Contents

I.2 Separation System Considerations

R. Wittkowski, Berlin

II Analysis of Food Ingredients

II.1 Components in Foods with a High Carbohydrate Content

O. Fröhlich, Würzburg

II.2   Fats and Compounds in Fat Containing Foods
O. Fröhlich, Würzburg

II.3   Aroma Compounds
G. Takeoka, Albany

II.4   Stereodifferentiation of Chiral Flavor and Aroma Compounds

K.-H. Engel, Berlin

II.5   High Molecular Weight Compounds

R. Hardt, Berlin

III   Analysis of Residues and Contaminants

III.1   Pesticide Analysis

C. Fürst, Münster

III.2 Contaminant Analysis
P. Fürst, Münster

III.3 Headspace Gas Chromatography of Highly Volatile Compounds
S. Vieths, Berlin

I Capillary Gas Chromatography

I.1 Fundamental Aspects of Capillary Gas Chromatography

R. Wittkowski

1 Theory of Gas Chromatography

1.1 The Chromatographic Process

1.1.1 The Distribution Constant K_D

As in all other chromatographic processes gas chromatography is based on the distribution of a component (solute) between two phases. When submitted to the column a solute immediately partitions between the stationary (liquid) and the mobile (gas) phase. The mobile phase (carrier gas) will then carry out the transport through the column after injection of a mixture of different components. Different solutes will interact differentially with the stationary phase depending on their molecular structure (Fig. I.1.–1).

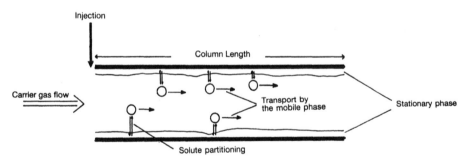

Fig. I.1.–1 Schematic of the chromatographic process.

This distribution is described with the distribution coefficient K_D, defined as the ratio of solute weight in equal volumes of the stationary and mobile phases:

$$K_D = \frac{\text{Solute weight per unit volume stationary phase}}{\text{Solute weight per unit volume mobile phase}} = \frac{C_s}{C_m} \tag{1}$$

K_D is a true equilibrium constant which depends only on the solute, the liquid phase, and the temperature. It is not dependent on the column type. The transport of the solutes through the column occurs only in the continuously moving gas phase with the same rate. The total time a solute remains in the column is its retention time (t_R). The longer a particular substance is in the gas phase the faster it is eluted from the column, which infers that t_R is small. Compounds which interact with the liquid phase to a greater extent remain on the column for a longer time and t_R therefore is larger. The interaction of a compound with the stationary phase, which actually determines the retention time depends on the molecular structure, especially on the type and number of functional groups present but also on the molecular geometry.

1.1.2   Partition Coefficient (Ratio) k and Phase Ratio β

When looking at the absolute amounts of solute in the gaseous and liquid phases instead of concentration the distribution coefficient K_D can be described as follows:

$$K_D = \frac{\text{amount of solute in liquid phase/volume of liquid phase}}{\text{amount of solute in gas phase/volume of gas phase}} \tag{2}$$

$$K_D = \frac{\text{amount of solute in liquid phase}}{\text{amount of solute in gas phase}} \times \frac{\text{volume of gas phase}}{\text{volume of liquid phase}} \tag{3}$$

$$K_D = k \cdot \beta \tag{4}$$

The phase ratio β is defined as the ratio between both the gas volume and the volume of the stationary phase within the column. For WCOT (wall coated open tubular) columns the following equation is applicable:

$$\beta = \frac{r}{2d_f} \tag{5}$$

where r is the radius of the column and d_f is the average thickness of the coating.

k is called the partition ratio or capacity ratio and is defined as the amount of solute in the stationary phase compared to the amount of that solute in the mobile phase. Consequently, k can be directly related to the retention time t_R (Equ. 6).

$$k = \frac{t_R - t_M}{t_M} = \frac{t'_R}{t_M} \tag{6}$$

1.1.3   The Retention Time

The entire time in which a particular component is in the column is termed the total retention time t_R. As already mentioned a solute spends a certain time in the liquid phase and the rest of the time in the gas phase. From the sum of these times the total retention time can be determined (Equ. 7).

$$t_R = t_M + t'_R \tag{7}$$

Generally, the transport of a solute within a column occurs only in the gas phase and then only exactly with the carrier gas flow velocity. From this follows that regardless of how often a solute interacts with the liquid phase the time it spends within the gas phase is constant. This is termed the "gas holdup time" (t_M). Therefore, the adjusted retention time $t'_R = t_R - t_M$. It can be determined directly by injecting substances, which do not interact with the stationary phase ($K_D \sim O$), as for example all noble gases. In practice methane is used, which essentially is not retained by the column. The relationship between these mathematical terms is shown in Fig. I.1.–2.

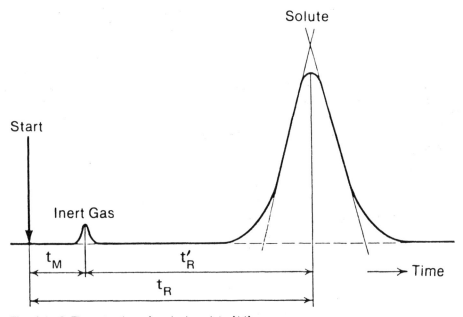

Fig. I.1.–2 The retention of a single solute [11].

The retention data of a substance are directly proportional to t'_R and therefore to k (Equ. 6).

1.2   Column Efficiency

1.2.1 Number of Theoretical Plates n

To obtain a good separation of substances it is necessary to keep the bandwidth of a particular substance as narrow as possible during its passage through the column. In addition, the standard variation of the retention time of identical molecules should be as small as possible. Even when identical molecules remain exactly the same time in the liquid phase, longitudinal diffusion (Equ. 33) will cause a certain peak broadening. Hence, the effectiveness of a separation of two components with different K_D values depends directly on peak broadening. In some unfortunate cases a larger retention time range for identical molecules can lead to overlapping of the tailing edge of the faster component with the leading edge of the slower component, and hence to an unsatisfactory separation. Thus, the separation efficiency describes the relationship between the retention time from the moment of injection (t_R) and the standard variation of peak broadening after emergence from the column (σ), and is defined as the number of theoretical plates, n:

$$n = \left(\frac{t_R}{\sigma}\right)^2 \tag{8}$$

This concept originated in the distillation terminology and is based on the assumption that a distribution process in equilibrium between the mobile and stationary phase in the column can be estimated by a number of individual hypothetical distribution steps (distillation steps). To circumvent a determination of σ, these peaks are assumed to have a Gaussian form, and n can then be determined by the easily measureable peak width at half height (2.354 σ) (Fig. I.1.–3).

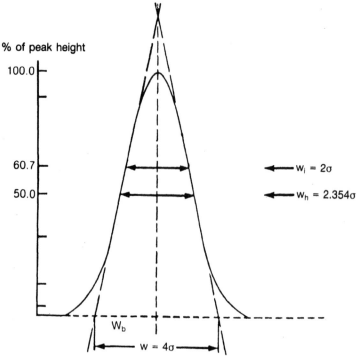

Fig. I.1.–3 Relationship between the measured peak width and σ of the Gaussian peak [11].

$W_{inflection}$ $\quad\quad = 2\sigma = W_i$ $\hfill$ (9)

$W_{half\ height}$ $\quad\quad = 2\sigma (2 \ln 2)^{1/2} = W_h$ $\hfill$ (10)

$\quad\quad\quad\quad\quad = 2.354\,\sigma$ $\hfill$ (11)

Equation (8) can therefore be described as follows:

$$n = \frac{(t_R)^2}{\left(\dfrac{W_h}{2.254}\right)} = 5.545 \left(\frac{t_R}{W_h}\right)^2 \tag{12}$$

or

$$n = \frac{(t_R)^2}{\left(\dfrac{W_h}{4}\right)} = 16 \left(\frac{t_R}{W_b}\right)^2 \tag{13}$$

1.2.2 Number of Effective Theoretical Plates N

The number of theoretical plates (n) relates to the total retention time, and thus to the sum of t'_R and t_M. With that t_M has also a positive influence on n even so it has no influence on the actual separation process. Hence, not all theoretical plates expressed by n are effective. For this reason PURNELL [1] and DESTY et al. [2] introduced the number of effective theoretical plates (N), which uses t'_R instead of t_R (Equ. 14).

$$N = 5.54 \left(\frac{t'_R}{W_h}\right) \tag{14}$$

Equations (15) and (16) show n and N in direct relationship, both are dependent on the capacity factor k. Therefore, k values should always be given together with the number of theoretical plates in order to obtain comparable values.

$$N = n \left(\frac{k}{k+1}\right)^2 \tag{15}$$

or

$$n = N \left(\frac{k+1}{k}\right)^2 \tag{16}$$

Figure I.1.–4 demonstrates that the difference between n and N becomes asymptotically smaller with increasing k. The reason for this is that with increasing total retention time the influence of t_M becomes smaller. It follows from that that the values of k and (k + 1) approach each other more and more until for $K = \infty$, n = N.

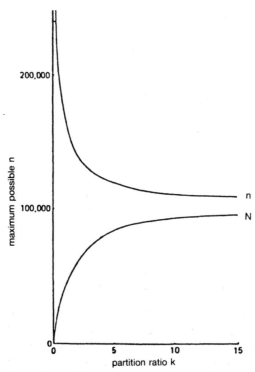

Fig. I.1.–4 Maximum number of theoretical plates (n) possible and effective theoretical plate numbers (N) calculated at μ_{opt} for each value of k.

In order to compare the separation capability of columns of different length the number of plates per unit length or vice versa the column length equivalent to one theoretical plate is often used (HETP, height equivalent to a theoretical plate, Equ. 17).

$$h = \frac{L}{n} \text{ (in mm)} \tag{17}$$

The height equivalent to an effective theoretical plate, HEETP, relates to that column portion which corresponds to one effective theoretical plate N (Equ. 18).

$$H = \frac{L}{N} \text{ (in mm)} \tag{18}$$

H and h decrease with increasing N and n, respectively.

1.2.3 Separation Number TZ

Another parameter to measure column efficiency is the separation number TZ (from the German word "Trennzahl" [3]) It is a measure of how many peaks with the same width at half height would theoretically fit between two real peaks in a homologous series containing x and (x + 1) carbon atoms (Equ. 19).

$$TZ = \frac{t_R(x+1) - t_R(x)}{W_h(x+1) + W_h(x)} - 1 \tag{19}$$

Normally, saturated straight chain hydrocarbons are used, however in very polar phases homologous alcohols or fatty acid methyl esters can also be used.

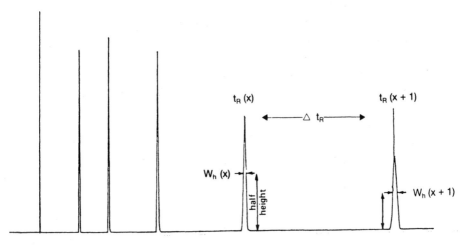

Fig. I.1.–5 Calculation of the Trennzahl TZ (separation number).

The separation number can be drawn upon isothermal as well as temperature programmed GC. It is a simple measurement to compare columns with similar phase coatings or to characterize the efficiency of a column over a longer period of time. Contrary to other descriptions of capillary separation efficiency the separation number is valid only for the particular capacity ratio and the specific temperature range, which it was calculated for. A determination of the separation number from a chromatogram is shown in Figure I.1.–5. According to Equation (20) the separation number is in direct relationship to resolution R:

$$TZ = \frac{R}{1.177} - 1 \tag{20}$$

This Equation assumes a resolution of R = 1.17 (4.6 σ) (compare Equs. 23–27). According to Equation (21) there is a relationship between the separation number and the effective theoretical plate number N when isothermal separations are performed.

$$TZ = 0.425 \frac{\alpha - 1}{\alpha + 1} \sqrt{N} - 1 \tag{21}$$

1.3 Separation of Components

1.3.1 Relative Retention α

In order to separate compounds their net retention times have to be different. The quotient between t_R of the later eluted substance (B) and t_R of the earlier eluted substance (A) is the relative retention time α (Equ. 22).

$$\alpha_{A,B} = \frac{t'_R (B)}{t'_R (A)} \qquad (22)$$

In the worst case α is equal to 1, but never smaller. Relative retention times vary with varying temperatures, because the influence of temperature increase is greater on the retention of the later eluting substance than on the earlier. In addition, the relative retention time α does not take into account that the quality of separation not only depends on the difference in retention time, but also on the peak broadening to a large extent.

Injection

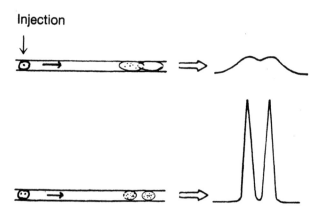

Fig. I.1.–6 Influence of the longitudinal diffusion on the separation of a substance pair in similar liquid phase and temperature, but different column efficiencies.

Figure I.1.–6 illustrates that even so the retention times for each of the components is equal in both cases the quality of separation in the lower chromatogram is significantly better simply due to the peak widths.

1.3.2 Resolution

An operand which takes into account both the separation according to retention as well as peak broadening is the resolution R (Equ. 23).

$$R = \frac{2 \left[t_R (B) - t_R (A) \right]}{W_b (B) + W_b (A)} \qquad (23)$$

As in the determination of n the peak width is measured preferably at half height instead of at the baseline, and it follows from that

$$R = \frac{2\,[\sim]}{1.699\,[W_h\,(B) + W_h\,(A)]} = \frac{1.177\,[\sim]}{[\quad\sim\quad]} = 1.177\,TZ + 1.177 \tag{24}$$

where

$$W_b = 4\,\sigma \text{ and } W_h \tag{25}$$

$$W_b = \left(\frac{4}{2.355}\right) W_h \tag{26}$$

$$W_b = 1.699 \quad W_h \tag{27}$$

(see also Fig. I.1.–3 and Equs. 9–11)

Usually in practice baseline separation is achieved at a resolution of about R = 1.5 (Fig. I.1.–7). In an ideal Gaussian distribution R = 1.0 would be sufficient.

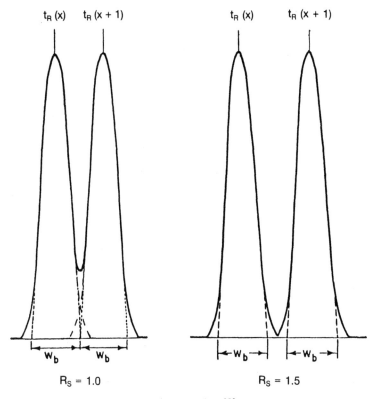

Fig. I.1.–7 Resolution and compound separation [8].

Problems always arise in asymmetric peak form or tailing. If the adjusted retention times of two separated compounds are known then R can be described as a function of the α-values (Equ. 28–30).

$$R = \frac{\sqrt{n}}{4} \left(\frac{\alpha - 1}{\alpha} \right) \left(\frac{k}{k+1} \right) \tag{28}$$

$$R = \sqrt{N} \left(\frac{\alpha - 1}{\alpha} \right), \text{ since } N = n \left(\frac{k}{k+1} \right)^2 \tag{29}$$

$$R = \frac{1}{4} \left(\frac{\alpha - 1}{\alpha} \right) \sqrt{\frac{L}{H}} \tag{30}$$

As can be seen from Equations (28) and (29) resolution is directly proportional to the square root of the number of theoretical plates, or, in the case of Equation (30) is directly proportional to the square root of the column length assuming α and H are kept constant, whereby it has to be considered that α is temperature dependent (see Chapter I.1.3.1.). Therefore, for a doubling of n at equal column efficiency the column length would have to be quadrupled. By changing Equation (28) into Equation (31) the number of theoretical plates for any desired resolution can be calculated.

$$\left(n_{req} = \right) 16 \, R^2 \left(\frac{\alpha}{\alpha - 1} \right)^2 \left(\frac{k+1}{k} \right)^2 \tag{31}$$

2 The Column

2.1 The VAN DEEMTER Equation

The separation of substances within a column is dependent on many parameters which include not only column parameters, but also the influence of carrier gas type and velocity as well. So, for example one can shorten the analysis time by increase in carrier gas velocity, but there is a point at which the efficiency of the column decreases due to a higher resistance to mass transfer resulting in peak broadening. When the velocity of the carrier gas is too low the influence of longitudinal diffusion becomes significant and leads to the same effect.

In 1956 VAN DEEMTER developed an equation in which the influences of different column parameters on carrier gas velocity is described [4]:

$$h = A + \frac{B}{\mu} + C \, \mu \tag{32}$$

The factors **A, B,** and **C** describe in simplified form the physical processes, which the solutes undergo during the chromatographic separation. **A** describes the influence of the multi flow path (Eddy-) diffusion on peak broadening. **B** describes the influence of the axial molecular diffusion in or against the direction of the carrier gas flow (longitudinal diffusion) on the peak broadening and is defined as:

$$B = 2 D_G \tag{33}$$

B has the dimension cm^2/s. D_G is the diffusion coefficient of the solute in the particular carrier gas. Normally **B** ranges in value between 0.01 and 1 cm^2/s. **C** describes that effect which due to the continuous transport of solutes by the carrier gas works against the complete adjustment of the distribution equilibrium K_D. According to the diffusion velocity in both phases the mass transfer between mobile and stationary phases is finite in relation to the carrier gas velocity. The **C**-term consists of the resistance to mass transfer in the liquid (C_L) and mobile phases (C_G) (Equ. 32). This differentiation is especially important when comparing capillary columns with packed columns, because in the latter the liquid film is significantly thicker and irregular and the solutes are separated by lower gas volumes.

$$C = C_G + C_L \tag{34}$$

$$\text{where} \quad C_G = \frac{r^2}{D_G} \cdot \frac{1 + 6k + 11 k^2}{24 (1 + k)^2} \tag{35}$$

$$\text{and} \quad C_L = \frac{d_f^2}{D_L} \cdot \frac{2 k^2}{4 (1 + k)^2} \tag{36}$$

In capillary columns the **A**-term is ineffective since due to the lack of column packing only one transport direction exists. The VAN DEEMTER Equation (32) therefore reduces to the GOLAY-Equation [5] (Equ. 37).

$$h = \frac{B}{\mu} + C \mu \tag{37}$$

In most chromatographic separations with capillaries the **C**-term is the dominant influence on peak broadening as will be shown in the following chapters.

2.2 Carrier Gas Velocity

The VAN DEEMTER Equation (Equ. 32) describes the effects of the carrier gas velocity on the efficiency of a column. Many chromatographers still measure the volumetric gas flow and neglect the fact that the solutes are transported through the column by the linear velocity of the carrier gas rather than the volumetric flow. Therefore, one should measure the flow as average linear gas velocity in cm/s rather than in ml/min. When F is the volumetric flow in ml/min one can with the help of Equations (38) and (39) calculate the flow in cm/s.

$$F = 0.6 \, \pi \, r^2 \, \mu \tag{38}$$

$$\mu = \frac{1.67 \, F}{\pi \, r^2} \tag{39}$$

r describes the column radius and μ is the average linear gas velocity. In contrast to liquids gases increase in viscosity with increasing temperature. For example H_2, N_2, and He show a relative viscosity increase of 40 % with a change in temperature from 40 to 260 °C [6]. Because most GC systems are pressure regulated rather than flow regulated a decrease in carrier gas flow occurs during a temperature programmed separation. Hence, the gas flow should be optimized for that temperature range of the column in which the most critical separation has to be accomplished.

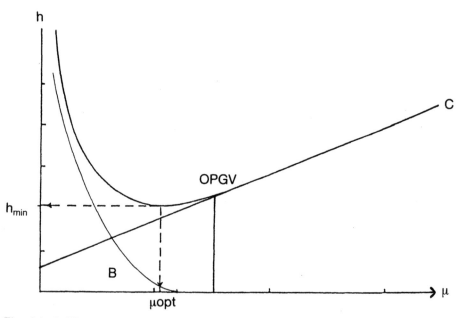

Fig. I.1.–8 Effect of the average linear carrier gas velocity μ on column efficiency h as determined by the longitudinal diffusion **B** and the resistance to mass transfer **C** of the van Deemter equation.

Figure I.1.–8 demonstrates the dependence of plate height h from the linear carrier gas velocity μ. This dependence is the summation of two opposing effects expressed by **B** and **C** (Equ. 37). The Eddy-diffusion **A** for capillary columns is zero. The peak broadening during low carrier gas velocities results mainly from the diffusion in the direction of the column axis (B), while at high carrier gas velocities only the resistance to mass transfer (C) is of importance. For low carrier gas velocities the latter term is small (small h, large n), since the substances have a greater chance to dissolve into and vaporize from the liquid phase. At higher carrier gas velocities, however, solutes contact less stationary phase per unit column length, which results in less distribution steps. At high carrier gas velocity the C-term is directly proportional to h (or the decrease of n). Theoretically, unfortunately only theoretically, the best results would be obtained whenever the longitudinal diffusion (B) is infinitely slow and the lateral diffusion would happen infinitely fast. Therefore the goal of the optimization is to adjust the carrier gas velocity μ in such a way that h is at minimum (h_{min}), i.e. n at maximum (h = L/n). Since n is a function of the square of the t_R/W_h ratio n will be decreased by any factors that make this ratio larger and increased by any factors that make the ratio smaller. These relationships are explained in Figure I.1.–9 in combination with Table I.1.–1.

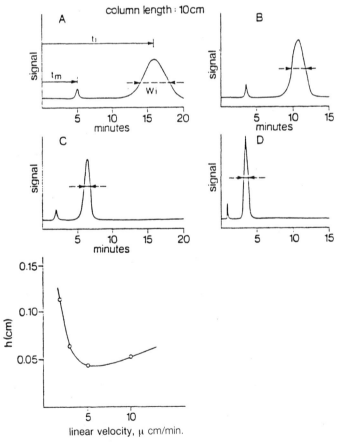

Fig. I.1.–9 Effect of μ on the shape of GC peaks [20].

Table I.1.–1 Data to the Corresponding Chromatograms in Figure I.1.–9 [20].

	A	B	C	D
t_M	5.0	3.5	2.0	1.0
μ	2	3	5	10
t_R	16.0	10.7	6.4	3.5
W_h	4	2	1	0.6
N	89	149	227	189
L	10	10	10	10
h	.113	.063	.044	.053

2.3 Optimum Practical Gas Velocity

In practice chromatographic separations are performed at higher carrier gas flows than μ_{opt} at the expense of separation efficiency. Especially in separation systems with flat VAN DEEMTER curves analysis time can be drastically shortened without great sacrifice of separation efficiency. This carrier gas velocity, which is termed the **o**ptimum **p**ractical **g**as velocity (OPGV), is localized at that point where the VAN DEEMTER curve changes into the linear region [7] (Fig. I.1.–8). However, experimental VAN DEEMTER plots do not exhibit linearity, because the only possibility to increase the average linear gas velocity is to change the pressure drop through the column, and thereby directly influence the velocity and diffusivity gradients that are engendered by the compressibility of the carrier gas. For this reason JENNINGS [8] suggested to better define the OPGV as that point, where the n/t_R ratio is maximized. These facts show that the calculated VAN DEEMTER curve is a theoretical curve containing many approximations. Nevertheless, it is useful because it allows the demonstration of many dependencies in gas chromatography.

2.4 Carrier Gas Selection

Basically, all inert gases can be used as carrier gases. Normally, however only nitrogen, helium, or hydrogen are used. Those three gases differ in their viscosity ($N_2 > He > H_2$), and therefore their influence on the diffusion coefficient (D_G) of the solute is different. D_G is a part of the B and C_G terms, in which it is inversely proportional and proportional to the linear gas velocity, respectively (Equs. 34–36).

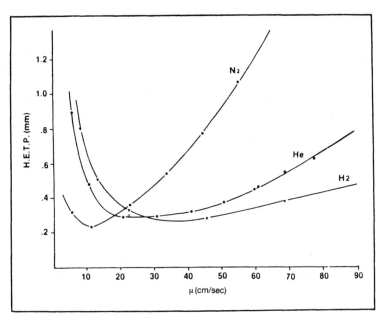

Fig. I.1.–10 Van Deemter curve for a 25 m × 0.25 mm WCOT column (d_f = 0.4 µm) for three mobile phases (N_2, He, H_2) [11].

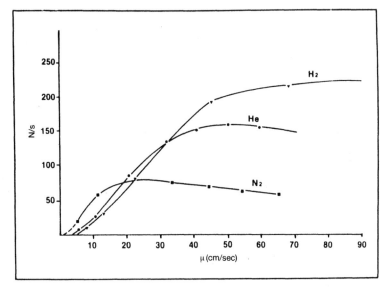

Fig. I.1.–11 Effective plates per second versus average linear gas velocity µ [11].

As an example Figure I.1.–10 depicts the VAN DEEMTER curves for the analysis of C_{17} at isothermal conditions (k = 4.95, WCOT OV-101, 25 m × 0.25 mm) in which all three carrier gases were employed. Lowest h_{min} was achieved with nitrogen. However, μ_{opt} for nitrogen is about 8–10 cm/s and h_{min} increases rapidly and n decreases when carrier gas velocity is increased resulting in rapid efficiency loss. For helium h_{min} is somewhat higher, on the other hand μ_{opt} is reached at higher values for μ and curve is flatter. Therefore, even carrier gas velocities > μ_{opt} are applicable without remarkable loss of separation power. The trend is most extreme when hydrogen is being used. With only a small sacrifice in h_{min}, μ_{opt} is now 35–40 cm/s resulting in a much shorter analysis time. Even a doubling of μ_{opt} results in hardly any loss in separation efficiency. Even more pointed is this effect when looking at the plot of μ versus number of effective theoretical plates/s (Fig. I.1.–11). This term characterizes the system performance which in routine analysis is controlled by analysis time and resolution.

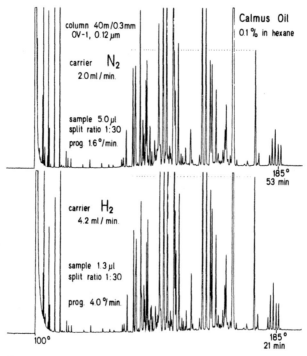

Fig. I.1.–12 Comparison of nitrogen and hydrogen as carrier gases for capillary columns.

For N_2 the conditions (temperature and flow rate) were selected for optimum resolution. For H_2 the conditions were set to yield identical elution temperature as for N_2, and for high but not maximum resolution [9].

It can be seen in Figure I.1.–11 that hydrogen delivers about four times as many effective plates/s as does nitrogen. With the help of the example of the analysis of calmus oil this effect [9] can be demonstrated (Fig. I.1.–12).

2.5   Column Parameters Influencing Efficiency

2.5.1   Film Thickness

The development of chemically bound phases in capillary GC has led to a wide spectrum in film thickness ranging from 0.05 to 8 µm. As already mentioned the phase ratio β is inversely proportional to the partition ratio k (Equ. 4). This means that whenever k becomes larger β becomes smaller and vice versa. Since β = $r/2d_f$, k increases with decreasing radius and increasing film thickness. k decreases with increasing r and decreasing d_f. Thus, retention time depends on the internal column diameter as well as on the film thickness. Therefore thin film capillary columns (d_f = 0.1 µm) are preferred whenever high boiling solutes such as polycyclic aromatic hydrocarbons (e.g. coronen, boiling point 525 °C), dioxins, etc. are to be analyzed. Thin films are thermally more stable, however they have only a small sample capacity so that low boiling compounds (C < 5) are not well separated. On the other hand, thick films are favored whenever low boiling materials are to be separated. The phase ratio decreases by a factor of 10 to 50 compared to thin film capillaries which results in a large increase in the capacity ratio for low boiling compounds. In this case the sample capacity approaches that of packed columns. Logically, they are favored in headspace analysis and trapping. In these particular cases the lower temperature stability as well as loss in sensitivity caused by an increase in peak broadening are not significant when compared to thin film capillaries. The influence of film thickness is shown in an example from GROB and GROB [9] in Figure I.1.–13.

The GOLAY Equation (37) in combination with Equation (36) make clear that the square of the film thickness is directly proportional to the resistance to mass transfer in the liquid phase. Therefore, when working with thin films (high β values) the influence of C_L on h and μ_{opt} can be ignored. Equation (40) in combination with Equations (33) and (35) make evident that h is independent from the diffusion coefficient in the gas phase (D_G).

$$h = 2 \sqrt{B \cdot C} = r \sqrt{\frac{1 + 6k + 11 k^2}{3 (1 + k)^2}} \qquad (40)$$

It follows from this that h_{min} for all three carrier gases at their respective μ_{opt} is nearly equal. On the other hand it follows from Equation (41) that μ_{opt} is directly proportional to the diffusivity in the gas phase (C_G).

$$\mu_{opt} = \sqrt{\frac{B}{C_G}} \qquad (41)$$

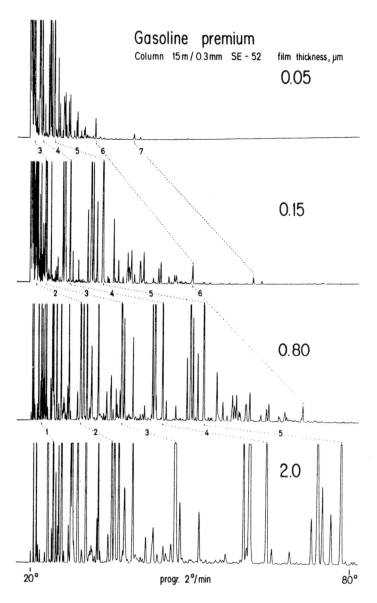

Fig. I.1.–13 Analysis of gasoline on columns with identical geometry and liquid phase, but different film thickness, under identical conditions.

1 n-pentane, **2** benzene, **3** toluene, **4** o-xylene, **5** 1, 2, 4-trimethylbenzene, **6** naphthalene, **7** 2-methylnaphthalene [9].

Since D_G is roughly inversely proportional to the square root of the molecular weight of the carrier gas it follows: μ_{opt} (H_2) = 1.4 μ_{opt} (He) = 3.7 μ_{opt} (N_2). In this case hydrogen should be used as carrier gas, because the retention time compared to nitrogen reduces by a factor of 3.7 while sensitivity is increased by the same order. This effect is elucidated by the curves in Figure I.1.–14 [10].

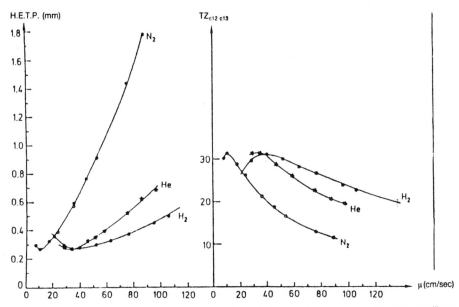

Fig. I.1.–14 h-μ and TZ-μ curves as function of the carrier gas selection for capillary columns with large β values. *Column:* 20 m × 0.27 mm OV-1, d$_f$ 0.22 μm [10].

The VAN DEEMTER curves of thick film capillaries (low β values) exhibit basic differences from those shown in Figure I.1.–14. On the example of a column coated with 5 μm poly-dimethylsiloxane (i.d. 0.32 mm) the differences in h$_{min}$ for nitrogen and hydrogen can be seen as well as the differences in the steepnesses of the curves (Fig. I.1.–15). μ_{opt} is found at low carrier gas flows. It follows that C_G (proportional to r^2/D_G) becomes smaller since D_G becomes larger at lower carrier gas flows. Consequently, the resistance to mass transfer (Equ. 34) is dominated by the C_L term (dependent on D_L and d$_f$). h$_{min}$ is much higher than in case of thin film columns. With increasing k value the value of h$_{min}$ decreases. Hence, in thick film capillaries carrier gases with higher viscosities are to be favored, because the exchange velocity between stationary and mobile phases is decreased significantly.

Only if analysis speed is of primary importance and the efficiency is sufficient for the particular separation problem hydrogen should still be used.

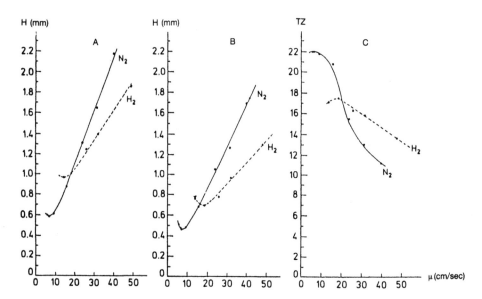

Fig. I.1.–15 h-μ and TZ-μ curves as function of the carrier gas selection for capillary columns with small β values.
A h-μ plot for k = 10, **B** h-μ plot for k = 20, **C** TZ-μ plots for C_8-C_9 [10].

2.5.2 Internal Diameter

The sample capacity (Equ. 44) and the separation efficiency (Equ. 40) of a capillary column depend directly on the inner diameter, and thereby the radius. SANDRA et. al. [10] produced five columns with differing radii, in which the coating in each case was equal and such that all columns had equal phase ratio β. With these the influence of column diameter (at large β values) on efficiency was investigated (Fig. I.1.–16).

Figure I.1.–16 and Table I.1.–2 indicate that for columns with large β values h_{min} is proportional to the inner diameter of the column. This can also be derived theoretically. According to Equation (40) for different values of k the following h_{min} values are determined:

$$\left.\begin{array}{l} \text{for k =} \quad 10, \ h_{min} = 1.79 \ r \\ \text{for k =} \quad 100, \ h_{min} = 1.90 \ r \\ \text{for k = } 1000, \ h_{min} = 1.91 \ r \end{array}\right\} \quad \sim 2 \ r$$

With increasing k value the value of h_{min} approaches that of the inner column diameter.

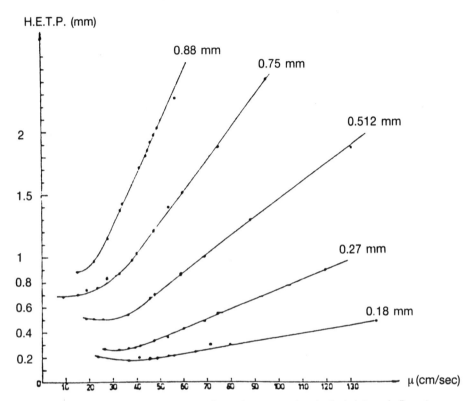

Fig. I.1.–16 Experimental h-μ curves for columns varying in their internal diameters. h was calculated for C_{12} at 100 °C with hydrogen as carrier gas [10].

Table I.1.–2 shows the corresponding data for the utilized columns.

Table I.1.–2 Data for Columns in Figure I.1.–16 [10].

ID (mm)	L (m)	$TZ_{C12-C13}$	TZ/m	HETP$_{min}$ Exp. (mm)	HETP$_{min}$ Theor. (mm)	μ (cm/s)	CE (%)	d_f (μm)
0.18	37	54	1.45	0.19	0.178	45	94	0.15
0.27	60	53	0.88	0.27	0.24	32	89	0.23
0.51	58	35	0.60	0.49	0.46	25	93	0.43
0.70	50	30	0.60	0.71	0.66	16	86	0.59
0.88	49	26	0.53	0.88	0.78	14	88	0.74

This fact can be used to estimate the theoretical number of plates n per meter (Equ. 42).

$$n = k/h_{min}, \text{ since } h_{min} \sim 2r \text{ it follows } n = k/2r \tag{42}$$

With a decrease of the inside diameter from 0.5 to 0.1 mm the number of theoretical plates per meter would increase from 200 to 1000 for k = 100. The optimal flow velocity μ_{opt} increases inversely proportional to the inner diameter, in this case by a factor of 5. For this the carrier gas pre-pressure would have to be increased significantly. At the same time however, the sample capacity would be reduced by a factor of $5^3 = 125$ (Equ. 44). These interrelationships make clear that there are limits in the optimization of separation efficiency by internal diameter reduction. The decrease in efficiency with increasing internal diameter (Fig. I.1.–16) can be compensated by increasing the column length by the same factor (n = L/h). In the case of thick film capillaries the influence of the internal diameter on h_{min} is minimized.

2.5.3 Column Length

The length of a column has a smaller influence on column efficiency than is normally expected. It follows from Equation (30) that the resolution is proportional to the square root of the column length. Therefore, in order to double the resolution the column length would have to be quadrupled. An increase in column efficiency therefore should first be attempted by decrease in internal diameter or change in film thickness. Also an optimization of column temperature can achieve this goal.

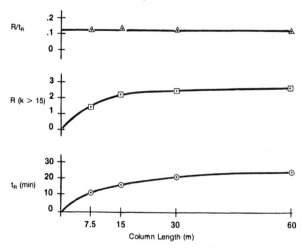

Fig. I.1.–17 Variation of resolution with column length.
Resolution between phenanthrene and anthracene as well as retention time of anthrancene is indicated.
(SE-52, d_f = 0.4 µm, µ = 40 cm/s H_2, 80 to 240 ° at 5 °C/min) [11].

Figure I.1–17 shows the change in resolution between phenanthrene and anthracene in relationship to column length. The most effective column length is between 15 and 20 meters, while resolution is not improved by a further increase in column length. The reason for this is the temperature programming effect. By increasing the temperature the solute is "reinjected" onto each successive increment of column length, until a point is reached, where the solute is effectively unretained [11].

2.5.4 Column Temperature

In general, the temperature has a very dominant influence on the efficiency of a column. The capacity ratio k is inversely proportional to column temperature. The lower the temperature the higher the value of k which leads to longer retention times. The strong influence of temperature programming on column efficiency is shown in Figure I.1.—18 [12].

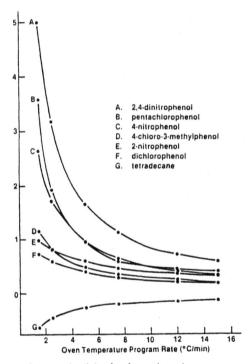

A. 2,4-dinitrophenol
B. pentachlorophenol
C. 4-nitrophenol
D. 4-chloro-3-methylphenol
E. 2-nitrophenol
F. dichlorophenol
G. tetradecane

Oven Temperature Program Rate (°C/min)

Fig. I.1.—18 Stationary phase selectivity for free phenols.
The greater the differences between the retention times of the phenol on the two stationary phases the greater the selectivity of the active phase for the particular phenol.
Columns: 30 m × 0.25 mm SE-30 and SE-54, μ = 50 cm/s H$_2$, temp 1 (2 min) = 75 °C [11].

Here the retention time differences of a phenol test mix analyzed on a SE-54 and a SE-30 column at different temperature program rates were investigated. At high program rates the retention time difference between both phases is small, that means the selectivity of the SE-54 column is lost. However, the selectivity of each individual phenol increases on the SE-54 phase when the program rate is reduced. This increase is so sufficient that here even a shorter column could have provided enough resolution. When selecting the temperature it should always be considered that during temperature increase the viscosity of the carrier gas increases and the velocity consequently decreases (Fig. I.1.–19).

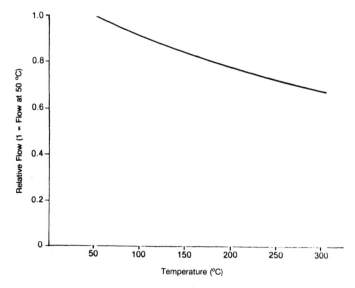

Fig. I.1.–19 Dependence of gas flow of helium (normalized to 50° C) on column temperature [11].

In general, most capillary columns are pressure regulated rather than flow regulated. Therefore, a temperature programmed analysis should be started at a higher carrier gas flow than would be used in an isothermal run. Since it has a flatter VAN DEEMTER curve, hydrogen has a great advantage over nitrogen or helium since the number of theoretical plates n varies less with increasing gas flow.

2.6 Coating Efficiency

In order to express the quality of a column coating in terms of numbers the ratio between the theoretical h_{min} and the experimental h_{min} which can be determined for every capillary column according to Equations (12) and (17), is calculated. The theoretical h_{min} is calculated according to Equation (40). From these values the coating efficiency can be determined according to Equation (43).

$$CE \% = \frac{h_{min} \text{(theor.)}}{h_{min} \text{(exp.)}} \cdot 100 = \frac{n \text{(exp)}}{n \text{(theor.)}} \cdot 100 \qquad (43)$$

2.7 Sample Capacity

The sample capacity of a capillary column depends on the volume of the stationary phase, and thus on the film thickness and internal column diameter. At sufficient solubility of the separating sample the sample capacity can be estimated according to Equation (44) and can be expressed as loading capability B in gram:

$$B = 0.05 \, M \cdot d^3 \cdot (1 + k) \cdot 10^{-6} \qquad (44)$$

where M is the molecular weight of the compound, d the internal diameter of the column in mm, and k the capacity ratio of the eluted compound. It becomes clear that a reduction of the internal capillary diameter by a factor of 2 will reduce the sample capacity by a factor of about 8. The influence of a reduction in film thickness (expressed indirectly through k) is by comparison smaller.

At lower solubility of the solutes in the stationary phase Equation (44) does not hold. Therefore, overloading effects can be observed in some peaks, whereas other substances in similar concentrations but differing polarity can exhibit symmetrical peaks.

2.8 The Stationary Phase

The development of stationary phases, column coating, and deactivating techniques, immobilisations, cross-linking, and other subjects which are not immediately connected with laboratory analysis are not the main subject of the following. Considerable work has been done in this field by K. Grob, G. Grob, and K. Grob Jr. as well as groups headed by L. Blomberg, W. G. Jennings, M. L. Lee, S. Lipsky, P. Sandra, and G. Schomburg among others. The interested reader is referred to these authors for further information.

Three types of solute-stationary phase interactions are important in gas chromatography. First is the dispersion interaction, which describes the state of intermolecular attractions caused by electron oscillations of an atom or molecule leading to electrical asymmetry. Second are dipole interactions occurring when both the stationary phase as well as the solute contain permanent dipole moments. Third are the base-acid interactions, the most important of which is the formation of hydrogen bonds of hydroxyl-containing solutes with the stationary phase. These interactions have the effect that polar compounds remain a longer period of time in polar phases and apolar solutes do the same in apolar phases. One talks about polar phases whenever they contain polar functional groups such as –CN, –CO, –OH, –C–O–, etc. [13]. Apolar phases are hydrocarbon-type such as Apiezon or squalane and polyester-type such as silicones. Table I.1.–3 gives an overview to the most common phases, which are mainly on polysiloxane basis, and equivalent phases offered by different manufacturers.

The silicone gums (OV-1, SE-30) are linear dimethylpolysiloxanes with higher molecular weights than the silicone oils (OV-101, SF-96, SP 2100), which normally are liquid. DB-1 and Ultra-1 columns are equivalent but cross-linked. These silicone gums form very homogenous films, show very high efficiency and can be used up to 350 °C (silicone oils only to 250 °C). SE-52 (5 % phenyl) and SE-54 (5 % phenyl and 1 % vinyl) are also capable of tolerating long-term exposures to 350 °C. The low low-temperature limit of −60 °C is also of advantage especially in the analysis of extremely volatile compounds, because these phases stay liquid even at these low temperatures, and thereby do not reduce their separating capabilities. This even allows cryofocussing on the column. The separation of solutes on methylsilicone phases is simply based on the sequence of boiling points, because of lack of any functional groups other interactions do not occur. To raise the selectivity other functional groups are incorporated which lean towards dipole and/or base-acid interactions. The maximum operating temperature, however, sinks with increasing numbers of functional groups to a lower limit of for instance 275 °C at 100 % cyanopropyl silicone (SP 2340, Silar 10CP).

In addition to the columns shown in Table I.1.−3 a large number of modified phases were developed for the solution of special problems. Especially mixtures with different contents of methylsilicone-, phenyl-, and cyanopropyl-phases increase the spectrum of conventional phases. So, for example, OV-275 contains 70−80 % CNPr, also Supelco offers a number of different columns with different CNPr contents as well: SP 2300 consists of 50 % CNPr and 50 % P, SP 2330 of 75 % CNPr and 25 % MS. DB-2330 (J&W Scientific), RT$_x$-2330 (Restec), and Silar 9CP (Chrompak) are similar.

100 % CNPr columns are also available such as SP 2340 (Supelco) and Silar-10CP (Chrompak). Phases consisting of mixtures of MS and P are offered by Restec under the trade names Rt$_x$-20 (80 % MS and 20 % P) and Rt$_x$-35 (65 % MS and 35 % P).

In the scope of increasing standardization of analytical methods, particularly the analysis of contaminants, specific GC phases were developed which show optimal separation properties for these particular materials. In this regard DB-608 (J&W Scientific) was produced for the analysis of pesticides and PCB's, and DB-624 (J&W Scientific) designed for the determination of purgeable hydrocarbons (EPA methods 601, 602, and 624). These, as well as compatible columns (Rt$_x$-Volatiles, Restec) are inert against water and methanol and are preferred in purge and trap analysis.

The DB-1301 column (J&W Scientific) is able to change selectivity and even elution order by varying the temperature program. The actual polarity of this particular column lies between that of SE-54 and OV-17, and has proven to be ideally suited for the determination of halocarbons of low and intermediary polarity [14, 15].

Table I.1.–3 Overview to the most common stationary phases and some important suppliers

Phase	Phase composition	Phases of similar polarity	J&W Sci.	HP	Quadrex	Chrompak	Supelco	SGE	RSL	Restec
SE-30, OV-1, OV-101	100 % MS	SP 2100, CP-SIL 5CP, UCW 982, SF 96, GB-1 SB-methyl-100, RSL-160, DL-200	DB-1	Ultra-1	007-1	CP-SIL 1CB	SPB-1	BP-1	150	Rt$_x$-1
SE-52, OV-73, OV-79, SE 54	95 % MS + 5 % P / 94 % MS + 5 % P + 1 % V	Dexsil 300, DC-200, DC-560, SB-phenyl-5, GB-5, Fluorolube, OV-3	DB-5	Ultra-2	007-2	CP-SIL 8CB	SPB-5	BP-5	200	Rt$_x$-5
OV-17	50 % MS + 50 % P	OV-11, DC-710, OV-22, GB-17, SB-phenyl-50, OV-25	DB-17	HP-17	007-17		SP 2250		300	Rt$_x$-50
OV-210	50 % MS + 50 % TFP	SP 2401, QF-1, UCON HB 280X Triton X-100	DB-210						400	
OV-225	50 % MS + 25 % P + 25 % CNPr	CS-5, UCON HB 5100, SB-cyanopropyl-25 AN-600, XE-60	DB-225	HP-225	007-225	CP-SIL 43CB		BP-15	500	Rt$_x$-225
DB-1301	94 % MS + 3% P + 3 % CNPr	–	DB-1301							
OV-1701	88 % MS + 6% P + 6 % CNPr	–	DB-1701		007-1701	CP-SIL 19CB		BP-10		Rt$_x$-1701
PEG	Polyethyleneglycol	Carbowax. DEGS, SP 1000, AT-1000, Superox, GB-20M	DB-Wax		007-20M	CP-Wax 52CB	Supelco-wax	BP-20		Stabil-wax^{TM}
OV-351	Nitroterephthalate modified PEG	FFAP, SP 1000								

MS = Methylsilicone   V = Vinyl   TFP = Trifluoropropyl
P = Phenyl   CNPr = Cyanopropyl   PEG = Polyethyleneglycol

The PEG phase (Carbowax) which has been of very frequent use in food analysis for many years is becoming less important in view of the development of more selective, temperature stable, and long-lived columns. The reasons for this are primarily its high sensitivity towards oxygen [16] and its solubility in water and low-molecular weight alcohols in addition to its relatively high low-temperature limit, which prevents separations below 50 °C. In the range of 50–60 °C PEG solidifies to a waxy solid. With the aid of multivariance analysis the properties of a number of different stationary phases can be compared (Fig. I.1.–20). The two-dimensional graph developed by STARK et al. [17] is based on a polarity scale and a value which is calculated from the retention of hydrocarbons caused by dipole-dipole interactions as well as base-acid interactions of alcohols.

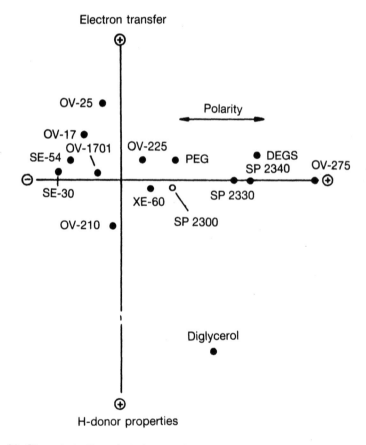

Fig. I.1.–20 Characterization of stationary phases by means of different possible interactions with the solutes.

2.9   Stationary Phase Selection

The chromatographer can become quite confused with the wide spectrum and ever increasing number of new developments of phases, which are often optimized for the separation of very specific compound pairs. It is common that columns containing stationary phase material of a specific type are not necessarily equivalent when produced by different manufacturers [8].

In addition, column designations are often unclear. In most cases, however, the number of columns for a specific laboratory can be reduced to a few to solve 80–90 % of all separation problems. In order to accomplish this, one has to first of all understand the compounds to be separated, i.e. are they polar or not, are they easily vaporizable, are they stable or thermolabile etc. Moreover, the presence of polar groups in the molecule must be clarified. In the analysis of chlorinated congeners such as dioxins and PCB's in particular one has to investigate whether or not a group specific analysis for the degree of chlorination is sufficient or if an isomer specific separation is necessarily required.

As a rule of thumb two apolar columns (OV-1, SE-54 or equivalent), one PEG-column as well as two additional columns (OV-1701, SP 2340) are sufficient in most cases. Apolar columns have the highest thermal stability and are always useful whenever screening analyses of unknown complex mixtures are to be performed, because they separate primarily according to the boiling point of the solutes which means that isomer groups are separated well, while individual isomers themselves are not particularly well separated.

The components of a sample are always eluted at a lower temperature from apolar than from polar columns, because the molecular interactions of apolar phases with the sample components is minimal. This leads to a reduction of the analysis temperature which in turn is of advantage in the analysis of thermolabile substances. Their chemical inertness also allows the analysis of samples with relative high water contents. For ultratrace analysis their low bleeding rates are beneficial. Chromatography on apolar columns, however, requires more care in the purification of the samples, since polar and high boiling impurities as well as strong acids and bases are able to shorten the life span of these columns considerably.

In general, polar phases allow a better separation of isomers even at minimal structural differences, and also tolerate extracts with polar impurities better. However, they are much more sensitive towards oxygen. The oxygen tolerance of stationary phases generally decreases in the following order: methyl polysiloxane > phenyl > cyanopropyl = trifluoropropyl >> PEG [8]. Also, traces of water reduce their life span considerably. Their thermal stability is lower, which hinders or even prevents the analysis of high boiling compounds. At temperatures below 50 °C analyses are limited since at these temperatures many polar phases solidify.

3 Retention Index System

The retention time of a substance is specific. Therefore, the simplest qualitative tool appears to be the comparison of retention data from known and unknown compounds. However, the multitude of materials which are analyzed today do not exclude coelution which can lead to misinterpretations. KOVATS [18] observed that if in a homologous series the logarithm of the adjusted retention times are plotted versus the carbon number a linear relationship is obtained. The KOVATS retention time index system bases on the expression of the retention time of a compound on a given stationary phase relative to the retention of normal paraffins which elute before and after the compound. By definition the retention index of the normal paraffins is 100n (n = carbon number). Thus, heptane has a retention index of 700 and octane a retention index of 800 on any liquid phase at any temperature. The retention index of any compound then is equal to 100-times the carbon number of a hypothetical normal paraffin exhibiting the same adjusted retention time as the substance of interest.

The retention index I of a compound A on liquid phase a at a temperature b is defined as

$$I_b^a = 100\,n + 100 \cdot \frac{(\log t'_R(A) - \log t'_R(N))}{(\log t'_R(N+1) - \log t'_R(N))} \tag{45}$$

where N is the number of carbons in the paraffin preceding compound A and (N + 1) is the carbon number of the hydrocarbon standard with retention time exceeding that of A.

To increase the certainty of the qualitative data obtained by GC at least two retention times on columns of different polarity should be obtained, for example on a dimethyl poly-siloxane stationary phase (OV-1, SE 30) and a PEG phase such as Carbowax 20M. Here one has to note however, that the retention data on PEG phases show greater variations (SD ± 0.5 − 1.0 %) than those measured on apolar phases (SD ± 0.1 %). The reason for this is that PEG phases are based on average molecular weights, and batch-to-batch variations in concentration and availability of functional groups are the rule rather than the exception [8].

VAN DEN DOOL and KRATZ [19] modified this system for the applicability of temperature pro-grammed GC analysis (Equ. 46).

$$I = 100\,n + 100_i \cdot \frac{t'_R(A) - t'_R(N)}{t'_R(N+1) - t'_R(N)} \tag{46}$$

where i is the carbon number difference of the corresponding hydrocarbons.

References

1. PURNELL, J. H. (1960) J. Chem. Soc., p. 1268

2. DESTY, D. H., GOLDUP, A., SWANTON, W. T. (1962) in "Gas Chromatography" (Brenner, N., Callen J. E, Weiss M. D., Eds.) Academic Press, New York, p. 105

3. KAISER, R. E. (1962) Z. Anal. Chem. **189**: 1

4. VAN DEEMTER, J. J., ZUIDERWEG, F. J., KLINGENBERG, A. (1956) Chem. Eng. Sci. **5**: 271

5. GOLAY, M. J. E. (1957) Nature **180**: 435

6. ETTRE, L. S. (1984) Chromatographia **18**: 234

7. SCOTT, R. P. W., HAZELDEAN, G. S. F. (1960) in "Gas Chromatography" (Scott, R. P. W., ED.) Butterworth, London, 144

8. JENNINGS, W. G. (1987) "Analytical Gas Chromatography", Academic Press Inc., Orlando, Florida 328867

9. GROB, K., GROB, G. (1979) J. High Resolut. Chromatogr. **2**: 109

10. SANDRA, P., PROOT, M., DIRICKS, G., DAVID, F. (1987) in "Capillary Gas Chromatography in Essential Oil Analysis" (Sandra, P., Bicchi, C., Eds.) Huethig Verlag, Heidelberg, p. 29

11. FREEMANN, R. R. (1981) "High Resolution Gas Chromatography", Hewlett-Packard Company

12. KEITH L. H. (Ed.) "Advances in the Identification and Analysis of Organic Pollutants in Water", Am. Arbor Science, Chapter 9

13. BURNS, W., HAWKES, S. J. (1977) J. Chromatogr. Sci. **15**: 185

14. GROB, K., GROB, G. (1983) Chromatographia **17**: 481

15. STARK, T. J., LARSON, P. A., DANDENEAU, R. D. (1983) Proc. 5th Int. Symp. Capillary Chrom., p. 65

16. EVANS, M. B. (1978) J. Chrom. **160**: 277

17. STARK, T. J., LARSON, P. A., DANDENEAU, R. D. (1983) J. Chromatogr. **279**: 31

18. KOVATS, E. (1958) Helv. Chim. Acta **41**: 1915

19. VAN DEN DOOL, H. H., KRATZ, P. (1963) J. Chromatogr. **11**: 463

20. KARASEK, F. W., CLEMENT, R. E. (1988) "Basic Chromatography—Mass Spectrometry", Elsevier, Amsterdam

I.2 Separation System Considerations

R. Wittkowski

1 Introduction

The motivations behind writing this article were not to deal with detailed theoretical fundamentals and general aspects of special instruments or instrumental parts such as injectors or detectors, developed for the exploitation of particular physico-chemical properties. Those basic considerations are better described in detail in special books and reviews, which the interested reader is kindly referred to [1, 4–12].

Alone, the gas chromatographic separation process itself can be manipulated in so many ways that it is worth it to highlight these advantages on the following pages.

Analyses in the field of food chemistry include almost all variations of handling chemical compositions and matrices. Hence, many different gas chromatographic methods have to be taken into consideration to cover the whole spectrum of possible combinations.

The application of capillary gas chromatography in food analysis basically falls into the following categories:

a. Separation of unknown constituents for identification
b. Differentiation of samples by generation of peak patterns for fingerprint comparisons
c. Separation and quantification of selected food ingredients
d. Separation and quantification of residues and contaminants
e. Food processing control
f. Separation for sensory evaluation

The variety of different interrelated subjects requires different instrument systems in terms of column design, injection techniques as well as sensitivity and selectivity of the detection mode. As far as specifically designed equipments are concerned, those are treated in the corresponding chapters of this book.

Chromatographic separations only give rise to the retentions of solutes, which can never prove what a solute is but only what it is not [1]. Therefore, it must be the aim of the analyst, to find a strategy to increase selectivity to guarantee a high degree of correspondence of peak- and substance identity. One strategy is to gain more information upon retention behavior, e. g. by determination of retention indices on a second column of dissimilar polarity, another includes the use of substance specific detection systems. Here, a series of common detectors are of great use, the most universal of which is the flame ionization detector (FID) because almost all volatile organic chemicals are detected with a response, which does not significantly depend on the chemical structure of the solutes. A likewise wide range of application is guaranteed by the thermal conductivity detector whereas its dynamic range (10^4) is smaller compared to that of the FID (10^7). Other detectors exhibit high sensitivity for special substance classes, whereas other compounds are not detected at all. The electron capture detector (ECD) is most specific to halogenated compounds and for nitrogen and phosphorus containing molecules the NPD (nitrogen phosphorus detector) had been designed. Moreover, a flame photometric detector was developed for the selective determination of sulfur containing compounds.

Much more structural information about the solutes can be obtained by directing the effluent of the capillary column to more complicated physico-chemical detectors. First of all mass spectrometry must be emphasized because of its great potential in terms of structure information. During the last two decades a broad spectrum of different ionization and detection techniques have been developed along with software assisted interpretation systems. A certain selectivity range to the analytes is given by different ionization modes. Electron impact ionization (EI) is applicable to produce a fragmentation pattern which is most helpful for structure elucidation. Positive (PCI) and negative chemical ionization (NCI) are gentle ionization techniques and therefore reduce the tendency to fragmentation with the benefit that sensitivity increases and the molecular ion appears more abundant. Depending on the problem to be solved either complete mass spectrometric fragmentation patterns are obtainable exhibiting a high degree of structural information or structure specific single ions are detected. Single (SID, SIM etc.) or multiple ion detection (MID) techniques permit higher sensitivities and are usually selected whenever well-known compounds are to be measured. They are applicable even if GC separations had been non-satisfactory for peak integration. Since mass spectrometers are very costly and personnel-intensive mass selective detectors (MSD, ITD) have captured the market and have prevailed in many fields of traditional GC and GC-MS domains.

Besides mass spectrometry also infrared spectroscopy gives access to valuable information concerning chemical structures. On-line connections of Fourier transform infrared spectroscopy (FTIR) with capillary gas chromatography are performed by a "light pipe" [2]. The compounds eluted from the column are non-destructively measured so that the effluent can be subsequently submitted to a mass spectrometer [3].

2 Coupling of Columns

Since the introduction of dead-volume-free coupling pieces a wide spectrum of combinatorial possibilities was opened to the chromatographer. Figure I.2.–1 shows three types of connectors using graphitized ferrules. Both upper connectors are used to couple fused silica capillaries of equal outer diameter. Here, difficulties arise if the column ends are not cut squarely or if pieces of silica or graphite fall into the inner area of the column by tightening the screws and twisting the column ends against each other. The lower construction depicted in Figure I.2.–1 had been designed for the analysis of PCDDs and PCDFs in the sub-ppt and ppq range including on-column injections of up to 20 μl. For this spe-

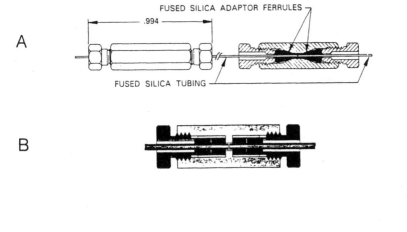

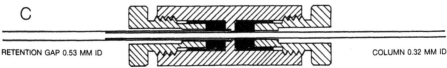

Fig. I.2.–1 Column connectors using graphitized ferrules (A: J&W Scientific, B: Gerstel GmbH, C: ref. 13, 14)

cial purpose the analytical column (0.32 mm i.d.) was inserted into a deactivated fused silica retention gap with an inner diameter of 0.53 mm (see also Chapter I.2.2.1). Although this is not a dead-volume-free connection the measurements never exhibited peak tailing [13, 14], because the amount of carrier gas per time unit trespassing both column sections is always equal. Consequently the carrier gas speed is accelerating towards the smaller diameter section leading to a suction effect.

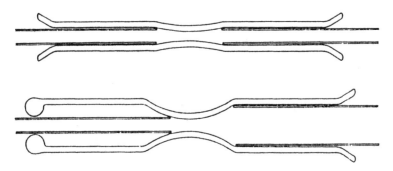

Fig. I.2.–2 All glass press-fit connector for fused silica columns

The introduction of so-called press-fit connectors a few years ago made it possible to perform column connections within a few seconds [15]. They consist of a glass tube with two conically formed ends (Fig. I.2.–2). Connectors of this type combine two advantages: zero dead-volume and totally inert all glass connections between fused silica columns. The outer polyimide coating of the fused silica column is acting as a ferrule when pressing the column ends into the connector.

Nevertheless after long time exposure to higher temperatures drawbacks in terms of leakiness may be observed.

2.1 Retention Gap

Whenever non-immobilized capillary columns are used, the injected solvent dissolves the stationary phase coating in the first column section, which is then transported further into the column. Hence, after several injections the first few meters of a capillary column are uncoated. Capillaries with immobilized and insoluble stationary phases, however, are stable in this regard. Here, the connection of the analytical column to a retention gap, which consists of an uncoated deactivated column segment is generally used to gain the same effect and is the most frequently used column combination. It protects the analytical column against pollution caused by non-volatile sample constituents. These are deposited in the retention gap, which can be easily exchanged from time to time [16–19]. Another beneficial effect is that peak distortion through the solvent injected can be circumvented. Depending on the volume injected the solvent immediately distributes over a certain column section moistening the column surface as a solvent film leading to a temporary dramatic increase of the capacity ratio and with that to two dissimilar effects:

1. Highly volatile compounds stay in the gas phase for instance after splitless injection and are retained at the capillary entrance because the solvent forms a film in the column after recondensation (Fig. I.2.–3A). For that the temperature of the capillary column, however, should be at least 20 to 40 °C below the boiling point of the solvent used [20–22]. At temperatures above the boiling point of the solvent the starting band grows broader and prevents a satisfactory peak resolution.

When the solvent is slowly evaporating the solutes get enriched at the end of the solvent film (Fig. I.2.–3B), get concentrated in a small band, and are finally transferred to the column with the rest of the solvent (Fig. I.2.–3C). This phenomenon is termed "the solvent effect" and is exploited for the enrichment of highly volatile compounds in GC analysis.

2. To high boiling solutes the recondensed solvent zone causes broadening of the starting band and thus leads to peak broadening and peak splitting. This effect is generally known as the "flooding effect". Figure I.2.–4 demonstrates, how far the breadth of the starting band of high boiling solutes can be narrowed by the installation of a retention gap.

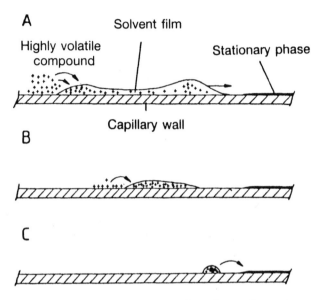

Fig. I.2.–3 Enrichment of highly volatile compounds at the capillary entrance by the solvent effect [4], (see text for details)

For the first moment after recondensation the solutes are irregularly distributed at the capillary entrance (A). After solvent evaporation the peak shapes are distorted (B). At the transition of the solutes from the uncoated to the coated section of the column they get refocused by the strong increase of the capacity ratio or through the dramatic decrease of the transportation speed (inversely proportional to $1 + k'$) (C, D) [23].

2.2   Effluent splitting

2.2.1   Dual Injection

Effluent splitting directly behind the injector offers the opportunity to transfer a sample to two different capillary columns with one single injection and thus enhance the informational value of an analysis because retention indices of two columns of differing polarity are obtainable. Such systems were at first only used including vaporizing (hot) injectors [24–27]. BICCHI et al. [28] used a system with a pre-column to enable on-column injections (Fig. I.2.–5) for analyzing volatiles of essential oils. A special variant of this technique is the splitting of the sample already in the injection syringe [29]. This system, however, requires two on-column injectors and offers a serious drawback because it is not applicable for automated analytical systems.

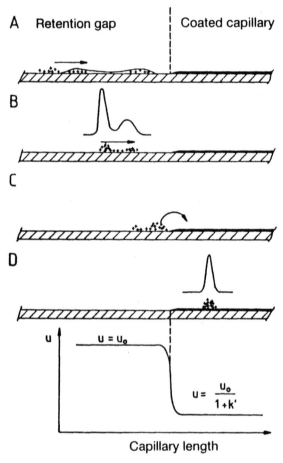

Fig. I.2.–4 Narrowing of the starting band of high boiling solutes through the installation of a retention gap [4], (see text for details)

The problem of automation of single injection/dual analyses involving on-column injection was circumvented by the introduction of the "glass cap-cross" inlet splitter [30, 31] (see also Fig. II.3.–4).

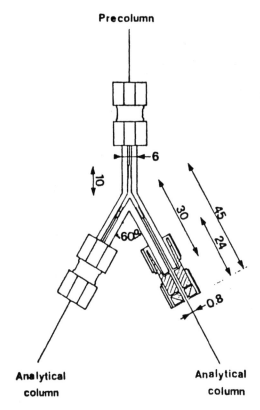

Fig. I.2.–5 Diagram of a splitting device [28]

2.2.2 Multidetection

On the other hand, effluent splitting at the end of a column prior to the detector enables simultaneous detection taking advantage of selectivities and sensitivities of different detection systems performing one injection. It has to be taken into account that the amounts of the detectable solutes are halved leading to a loss in sensitivity for each detector. Figure I.2.–6 clearly points to the differences in detector selectivity when polychlorinated biphenyls and hydrocarbons, the main constituents of waste oil, are measured simultaneously with an FID and N-FID [5]. Similar selective determinations can be achieved with the combination of FID and NPD as is depicted in Figure I.2.–7 [5]. Since the ECD is non-destructively operating the trespassing effluent stream can be additionally directed to another detector, e. g. an FID (Fig. I.2.–8). Thereby a further reduction of the analyte concentration is dropped. In this manner three chromatograms from one single injection are easily obtainable [32].

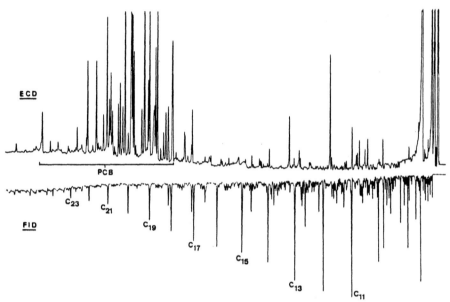

Fig. I.2.–6 Determination of trace amounts of isomeric polychlorinated biphenyls and hydrocarbons in a waste oil by parallel detection with FID and ECD [5]

3 Multidimensional Separations

Whenever multicomponent mixtures have to be analyzed, separation problems may occur, which cannot be solved by modifying temperature programming, changing film thickness or the polarity of the stationary phase or simply increase column length.

Those problems are better treated by techniques involving multiple column systems combined with multiple gas flow paths through the chromatographic separation systems or any other system enhancing information of one chromatographic analysis. These techniques are usually summarized by the term multidimensional chromatography. The term "two-dimensional chromatography" was used by BERTSCH [33] for a separating system containing two columns of different selectivities operated under conditions that eliminate ambiguities in correlating the solutes eluting from the two columns. Whenever peaks eluted from one column can be correlated with peaks from a second column, two-dimensional systems can be used to establish exact retention indices on two different stationary phases [1].

3.1 Stationary Phase Mixtures

The simplest way to separate solutes on two dissimilar stationary phases is the use of two columns connected in series using a press-fit connector (Fig. I.2.–2). In multicomponent

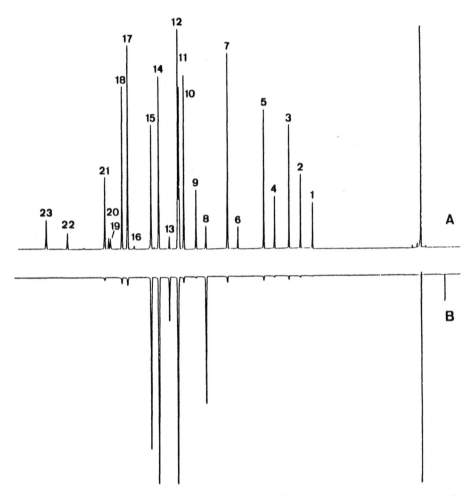

Fig. I.2.–7 Chromatograms of a test mixture including polyaromatic and heterocyclic hydrocarbons. Parallel detection with FID (A) and N–FID (B) [5]

1 Acenaphthene, **2** Acenaphthylene, **3** Dibenzofuran, **4** 2,3,6-Trimethylnaphthalene, **5** Fluorene, **6** Xanthene, **7** 9,10-Dihydroanthracene, **8** Phenacin, **9** Dibenzothiophene, **10** Phenanthrene, **11** Anthracene, **12** Acridin, **13** Carbazol, **14** Phenoxacin, **15** 9,10-Dihydroacridin, **16** 9-Methylphenanthrene, **17** 2-Methylphenanthrene, **18** 1-Methylphenanthrene, **19** Thianthrene, **20** 9-Methylanthracene, **21** 2-Phenylnaphthalene, **22** Fluoranthene, **23** Pyrene

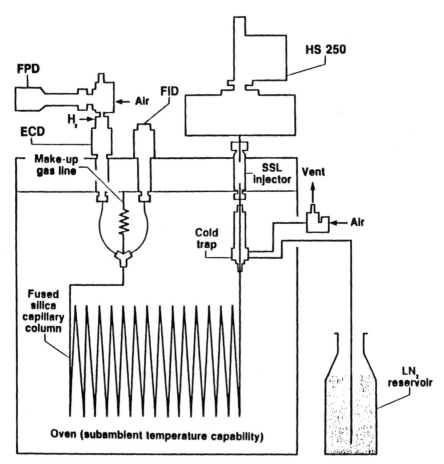

Fig. I.2.-8 Triple detection system connected to a headspace GC configuration with cold trapping device [32]

mixtures consisting of solutes of different chemical compositions, however, those systems rarely lead to satisfactory results, because critical separations from the first column may be optimized on the second column, but other solutes separated well on the first column may co-elute after trespassing the second column section. None the less selectivity can be increased in cases of specific separation problems and may serve as a potential tool in the separation of certain critical solute pairs.

The fundamentals of chromatographic separations using mixtures of stationary phases arose from investigations by MAIER and KARPATHY [34] who first employed those binary stationary phases. They calculated that the K_D, or k, if β is constant, of a solute in a binary mixture is a linear function of the volume fraction of either stationary phase:

$$K_{D(A + B)} = K_{D(A)} \, \Phi_{(A)} + K_{D(B)} \, \Phi_{(B)}$$

where A and B are two different stationary phases and $\Phi_{(A)}$ and $\Phi_{(B)}$ represent the volume fractions of A and B in the binary mixture. Later investigations by LAUB and PURNELL [35, 36] resulted in a more precise method to calculate the ideal phase mixtures. Instead of using the K_D values they used the relative retentions α of each solute pair and referred those to the volume percentage of one stationary phase in a binary mixture. In a graphical illustration of this relationship so-called window diagrams become visible as is exemplarily shown in Figure I.2.–9 for a hypothetical mixture of three components. There are two windows, one at 22.5 and one at 55 volume percent of stationary phase B in A [35, 36].

These calculations serve to optimize separations of special solute pairs. Once calculated the optimum, there are two possibilities to arrange phase mixing: the first is as already mentioned to couple two column sections directly (the percentages of A and B can be calculated from the column length and the film thickness) or secondly, to design a new column by mixing the two stationary phases directly before coating the column. In this manner the DB–1301 column was developed [37].

A remarkable example for the first application was demonstrated by TAKEOKA et al. [38]. Figure I.2.–10 shows the separation of a test mixture on a 30 m x 0.33 mm fused silica column with a 0.25 μm DB–1 coating. The total analysis time of 25 minutes did not ensure satisfactory separation. This, however, was achieved by use of a 30 m x 0.25 mm DB–1701 column with a total analysis time of 57 minutes (Fig. I.2.–11 A). After computer-calculated optimum phase combination of DB–1 and DB–1701 the tested compounds were separated within 2.9 minutes (Fig. I.2.–11 B). For that 2.9 m of the apolar column used in Figure I.2.–10 and 3.1 m of the DB–1701 column (Fig. I.2.–11 A) were connected and operated under optimum carrier gas velocity.

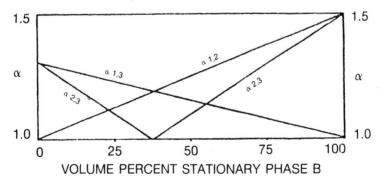

Fig. I.2.–9 Window diagram for a hypothetical mixture (ref. 35, 36, see text for details)

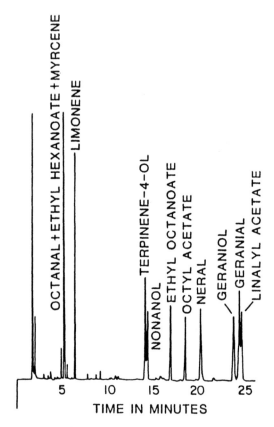

Fig. I.2.–10 Chromatogram of a test mixture separated on a 30 m x 0.33 mm DB–1 fused silica capillary [38]

3.2 Column Switching Systems

By addition of another carrier gas stream to systems of two or more capillary columns using a valve, the sample flow can be directed to different targets. The switching of the flow direction was performed by mechanical valves [39–41] as well as by pneumatical systems [24, 41–46]. Modern systems of multidimensional chromatography originate from the investigations of DEANS [45]. He introduced a pressure balancing system instead of a valve system to change the direction of the sample flow through a system of columns. Pressure switching systems offer lower dead-volumes and hence better chromatographic performance. Moreover, they do not limit the operating temperature and the solutes do not come into contact with the valve surfaces.

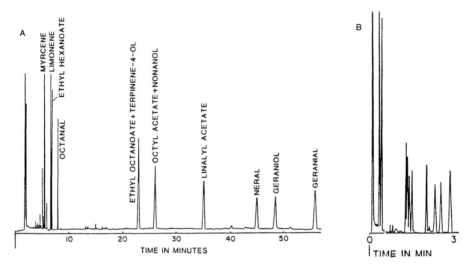

Fig. I.2.–11 Chromatogram of the test mixture separated on a 30 rn x 0.25 mm
DB-1701 column (left) and on a coupled column of 2.9 and 3.1 m seg-
ments (right) [38], (see text for details)

A so-called "Live"-piece is depicted in Figure I.2.–12 showing the flow directions at the
main pressure configurations. The benefits can be summarized as follows:

1. The analytical column can be protected from pollution with high boiling compounds,
which do not belong to the analytes of interest, by back-flushing them before they reach
the column. The back-flush system usually consists of a short pre-column and a T-piece
connection to the main-column. After producing a full chromatographic analysis the time
for passing the T-piece for the components of interest can be calculated. After that time
the carrier gas flow is directed through the T-piece and transports the wanted solutes
through the main column and the unwanted components through the pre-column out to
atmosphere (Fig. I.2.–12C). That shortens the analysis time, because all solutes eluting
after the requested compounds are not transported through the column, which additionally protects the analytical column from the overburden of unnecessary temperature
stress.

2. The main benefit of a pneumatically column switching system, however, is due to the
time-programmed switching of carrier gas flows. Insufficiently resolved peaks from the
first column are selectively transferred to a second column of differing polarity to obtain
satisfactory resolution. Thus, with the so-called "heart-cutting" every desired section of a
chromatogram can be cut out for further chromatographic operations (Fig. I.2.–12B).

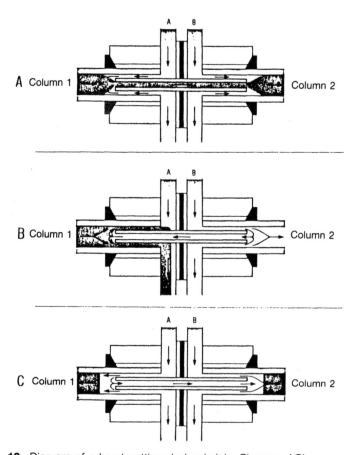

Fig. I.2.–12 Diagram of a heart-cutting device (origin: Siemens AG)

A: Straight operation

The pressure of control line B is lowered compared to the pressure of control line A so that all of the carrier gas flows out of the pre-column to the main-column.

B: Heart-cut

The pressure of control line A is slightly lowered compared to the pressure of control line B. The flow in the platinum–iridium capillary is opposite to that in the straight operation. The carrier gas flow does not go to the main detector but to the monitor detector.

C: Back-flush

The valve supplying column 1 with carrier gas is closed. The back flow in capillary operation goes through the split valve to a vent.

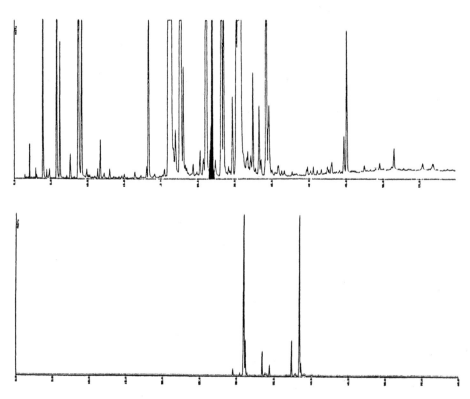

Fig. I.2.–13 Heart-cut transfer of a single peak of a peppermint oil from a polar pre-column (upper) to an apolar main-column (lower) [47]

Actually, this is the only chromatographic technique deserving the term "multidimensional". Other systems such as multidetection or single injection/dual analysis simply increase the reliability of a chromatographic result without exceeding the traditional separation dimension. The heart-cut system, however, introduces a second polarity to the separation system and thus another dimension. An example is given in Figure I.2.–13 showing the transfer of one peak of a peppermint oil extract from a polar pre-column to an apolar main-column. This example clearly demonstrates what retention behavior on one single stationary phase really means [47]. The same procedure was used by GOEKE-LER et al. [48] to quantify mono- and diethylene glycol in wine. In the middle of the eighties there had been public concern about the adulteration of Austrian wines with the latter compound.

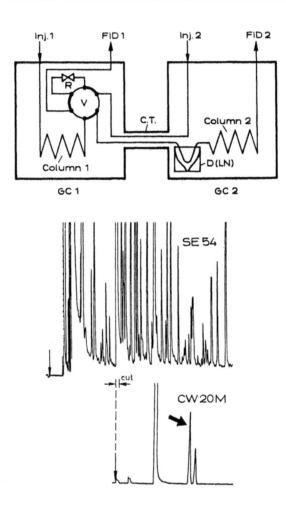

Fig. I.2.–14 Schematic diagram of a two-dimensional GC system and the two-dimensional separation of a gas chromatographic fraction from an extract of a fruit juice into one major component (butanoic acid methyl ester) and two minor compounds. The arrow denotes the peak with a typical odor.
GC1, GC2: individual GC's
C. T.: connection tube
V: micro-valve
R: restriction
D (LN): Dewar flask filled with liquid nitrogen
Inj. 1, 2: injectors
FID: flame ionization detector [49]

3. The heart-cut technique can also be successfully employed in samples where low concentrated peaks of interest are overwhelmed by very large peaks. From the large peak that section in which the trace analyte is expected is cut out leading to a much better resolution in the second column due to the favorable relative proportions.

The application of multidimensional gas chromatography with two column ovens in which each column can be operated independently at different temperature programs was used by ADAM et al. [49]. They successfully claimed to correlate the production of fruit juices with aroma changes (Fig. I.2.–14).

4. By inserting a cold trap (intermediate trapping) trace components are enriched because the pre-separation can be repeated for several times (Fig. I.2.–15). That enhances the signal to noise ratios of the analytes for the second column. Another field of application for intermediate trapping is in cases when the requested solutes must be analyzed in one chromatographic run but are eluted with very differing retention times from the first column. Also a hydrocarbon mixture can be injected after trapping the solutes of interest to enable the calculation of retention indices on the second column.

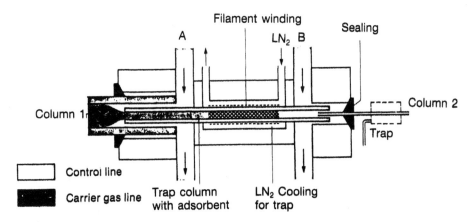

Fig. I.2.–15   Design of a coupling T-piece for the total transfer of a sample from column 1 to column 2 using an intermediate cold trap (origin: Siemens AG)

4   On-Line Coupled Liquid Chromatography-Gas Chromatography (LC-GC)

The potential of HPLC for sample clean-up for subsequent GC separations is well-accepted and often used off-line for sample pre-separations. However, the collection, concentration, and re-injection of the separate fractions when working off-line have the disadvantage that only a few percent of the eluted LC-fraction can be transferred to the GC column [50]. On-line coupling with conventional size HPLC columns have their drawback in only transferring a few microliters of the LC eluent to the GC system. This problem was solved by the introduction of long retention gap pre-columns capable of tolerating injections of several hundred microliters [51–53] but offered their disadvantage in periods of about an hour to evaporate the solvent [54]. The scheme of an instrumental device for such an on-line LC-GC system is shown in Figure I.2.–16 [55]. Some authors used packed fused silica micro LC columns [50, 56] and reached a transfer reproducibility of 3 percent at 7 percent relative standard deviation [50]. The only drawback of this technique is the low sample capacity but it has advantages in low eluent flow rates which allows the total transfer of the eluent to the GC and makes only few amounts of solvent necessary.

GROB JR. and STOLL [57] first introduced a loop-type interface for concurrent solvent evaporation for the on-line LC isolation and GC separation of raspberry ketone from a raspberry sauce and herewith offered one of the very few applications in the field of food analysis. Other authors used this system for the determination of biphenyls in fish [58] and diisooctyl phthalate in salad oil [59]. Papers written by other authors deal with environmental and clinical aspects [50, 60–62].

At least it should be mentioned that there are severe limitations in the use of polar solvents, e. g. in reversed phase LC, because they rapidly destroy the deactivation of the retention gaps.

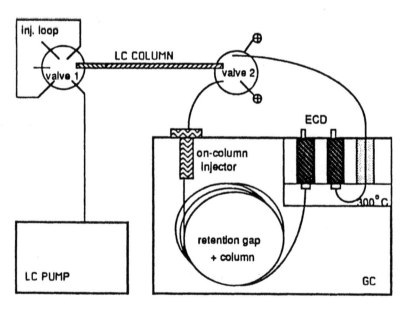

Fig. I.2.–16 Instrumental set-up for on-line LC–GC [55]

Nevertheless, the technique of on-line coupling of LC and GC combines two potential and efficient separation systems. It is a quite new technique and future developments with regard to easy handling and automation will determine at least the number of publications in the area of food chemistry.

5 Supercritical Fluid Chromatography (SFC)

Chromatographic separations with supercritical fluids as mobile phases were first performed by KLESPER et al. in 1962 [63].

Since the introduction of chemically bound stationary phases in capillary gas chromatography, SFC gained interest again. In one certain point the application of compressed gases in chromatographic separations is superior to classical GC techniques: fluid media offer a much better elution power. Compressed gases exhibiting diffusion coefficients in the range of 10^{-4} cm^2 per second are physically much more similar to fluids (10^{-5} cm^2/s) than to gases (10^{-1} cm^2/s). That means in other words that solutes can be separated at much lower temperatures, which consequently focuses the use of SFC to high boiling and thermolabile compounds. To speed up the elution the density of the mobile phase is programmable similar to the gradient elution in HPLC. Therefore, SFC is predicted to close the gap between capillary gas chromatography and high performance liquid chromatography and obliterates the former clearly defined limitations between those two analytical techniques.

When using CO_2, N_2 or SF_6 as mobile phases the employment of the universal and sensitive flame ionization detector (FID) is possible. With it the necessity of a chromophore, which is essential for a photometric detection in HPLC no longer exists.

Meanwhile, experts disagree about to what extent this new technique will contribute to the variety of analytical tools in food analysis. A final decision about the importance of SFC in food chemistry is not possible to this time but it is a great challenge to improve it for that reason.

Those readers who are interested in the theory and instrumental state are kindly referred to reviews on that topic [64–67].

References

1. JENNINGS, W. G. (1987) "Analytical Gas Chromatography", Academic Press Inc., Orlando, Florida

2. HERRES, W. (1987) "HRGC-FTIR: Capillary Gas Chromatography-Fourier Transform Infrared Spectroscopy", Huethig Verlag, Heidelberg

3. WILKINS, C. L. (1987) Anal. Chem. **59**: 571A–581A

4. OEHME, M. (1986) "Hochauflösende Gaschromatographie", Huethig Verlag, Heidelberg

5. SCHOMBURG, G. (1987) "Gaschromatographie", VCH Verlagsgesellschaft, Weinheim

6. MARKIDES, K. E., Lee, M. L. (1986) "Advances in Capillary Chromatography", Huethig Verlag, Heidelberg

7. GROB, K. "On-Column Injection in Capillary Gas Chromatography", Huethig Verlag, Heidelberg

8. FREEMAN, R. R. (1981) "High Resolution Gas Chromatography", Hewlett Packard Company

9. SANDRA, P. (1988) "Sample Introduction in Capillary Gas Chromatography", Huethig Verlag, Heidelberg

10. GROB, K. (1986) "Making and Manipulating Capillary Columns", Huethig Verlag, Heidelberg

11. JENNINGS, W. G. (1983) "Comparison of Fused Silica and Other Glass Columns in Gas Chromatography", Huethig Verlag, Heidelberg

12. SANDRA, P., BICCHI, C., (Eds.) (1987) "Capillary Gas Chromatography in Essential Oil Analysis", Huethig Verlag, Heidelberg

13. BECK, H., ECKART, K., MATHAR W., RÜHL, Ch.-S., WITTKOWSKI, R. (1988) Biomed. & Environm. Mass Spectrom. **16**: 161

14. BECK, H., ECKART, K., MATTHAR, W., WITTKOWSKI, R. (1988) Lebensmittelchem. Gerichtl. Chem. **42**: 101

15. ROHWER, E. R.., PRETORIUS, V., APPS, P. J. (1986) J. High Res. Chromatogr. **9**: 295

16. GROB Jr., K. (1982) J. Chromatogr. **237**: 15

17. GROB Jr., K., MÜLLER, R. (1982) J. Chromatogr. **244**: 185

18. GROB Jr., K., LÄUBLI, T. (1986) J. Chromatogr. **357**: 357

19. GROB. Jr., K., SCHILLING, B. (1987) J. Chromatogr. **391**: 3

20. GROB, K., GROB Jr., K. (1979) J. High Res. Chromatogr. **1**: 57

21. GROB Jr., K. (1982) J. Chromatogr. **17**: 253

22. GROB Jr., K., ROMANN, A. (1981) J. Chromatogr. **214**: 118

23. GROB Jr., K. KARRER, G., RIEKOLA, M.-L. (1984) J. Chromatogr. **334**: 129

24. SCHOMBURG, G. HUSMANN, H., WEEKE, F. (1975) J. Chromatogr. **112**: 205

25. KUGLER, E., HALANG, W., SCHLENKERMANN, R., WEBEL, H., LANGLAIS, R. (1977) Chromatographia **10**: 438

26. SANDRA, P., PROOT, M., DIRICKS, G., DAVID, F. (1987) in "Capillary Gas Chromatography in Essential Oil Analysis" (Sandra, P., Bicchi, C., Eds.) Huethig Verlag, Heidelberg

27. PHILLIPS, R. J. (1987) in "Capillary Gas Chromatography in Essential Oil Analysis" (Sandra, P., Bicchi, C., Eds.), Huethig Verlag, Heidelberg

28. BICCHI, C., FRATTINI, C., NANO, G. M., D'AMATO, A. (1987) Proc. 8th Int. Symp. on Capillary Chrom., Huethig Verlag, Heidelberg, p. 378

29. SEEKAMP, H., SANDRA, P., DAVID, F. (1987) Proc. 8th Int. Symp. on Capillary Chrom., Huethig Verlag, Heidelberg, p. 282

30. BRETSCHNEIDER, W., WERKHOFF, P. (1988) J. High Res. Chromatogr. **11**: 543

31. BRETSCHNEIDER, W., WERKHOFF, P. (1988) J. High Res. Chromatogr. **11**: 589

32. GAGLIARDI, P. (1987) Proc. 8th Int. Symp. on Capillary Chrom., Huethig Verlag Heidelberg, p. 390

33. BERTSCH, W. (1978) J. High Res. Chromatogr. **1**: 187, 289

34. MAIER, H. J., KARPATHY, O. C. (1962) J. Chromatogr. **8**: 308

35. LAUB, R. J., PURNELL, J. H. (1975) J. Chromatogr. **112**: 71

36. LAUB, R. J., PURNELL, J. H. (1976) Anal. Chem. **48**: 799

37. MEHRAN, M. F., COOPER, W. J., LAUTAMO, R., FREEMAN, R., JENNINGS, W. G. (1985) J. High Res. Chromatogr. **8**: 715

38. TAKEOKA, G., RICHARD, H. M., MEHRAN, M., JENNINGS, W. G., (1983) J. High Res. Chromatogr. **6**: 145

39. MILLER, R. J., STERNS, D., FREEMAN, R. R. (1979) J. High Res. Chromatogr. **2**: 55

40. JENNINGS, W. G. (1984) J. Chromatogr. Sci. **22**: 129

41. GORDON, B. M., RIX, C. E., BORGERDING, M. F. (1985) J. Chromatogr. Sci. **23**: 1

42. PHILLIPS, R. J., KNAUSS, K. A., FREEMAN, R. R., (1982) J. High Res. Chromatogr. **5**: 546

43. SCHOMBURG, G., WEEKE, F., MÜLLER, F., OREANS, M. (1983) Chromatographia **16**: 87

44. SCHOMBURG, G., BASTIAN, E., BEHLAU, H., HUSMANN, H., WEEKE, F. (1984) J. High Res. Chromatogr. **7**: 4

45. DEANS, D. R. (1965) J. Chromatogr. **18**: 477 **1**: 18

46. DEANS, D. R. (1968) Chromatographia **1**: 18

47. JOHNSON, G. L., TIPLER, A. (1987) Proc. 8th Int. Symp. on Capillary Chrom., Huethig Verlag, Heidelberg, p. 540

48. GÖKELER, U. K., MÜLLER, F. (1987) Proc. 8th Int. Symp. on Capillary Chrom., Huethig Verlag, Heidelberg, p. 518

49. ADAM, S. T. (1987) Proc. 8th Int. Symp. on Capillary Chrom., Huethig Verlag, Heidelberg, p. 574

50. DUQUET, D., DEWAELE, C., VERZELE, M. (1988) J. High Res. Chromatogr. **11**: 252

51. GROB, K., SCHILLING, B. (1984) J. High Res. Chromatogr. **7**: 531

52. GROB, Jr., K., NEUKOM, H. P. (1986) J. High Res. Chromatogr. **9**: 405

53. GROB, Jr., K., FRÖHLICH, D., SCHILLING, B., NEUKOM, H. P., NÄGELI, P. (1984) J. Chromatogr. **295**: 55

54. MUNARI, F., TRISCIANI, A., MAPELLI, G., TRESTIANU, S., GROB, Jr., K., COLIN, J. M. (1985) J. High Res. Chromatogr. **8**: 601

55. MARIS, F. A., NOROOZIAN, E., OTTEN, R. R., van DIJCK, R. C. J. M., de JONG, G. J., BRINKMAN, U. A. T. (1988) J. High Res. Chromatogr. **11**: 197

56. VERZELE, M., DEWAELE, C., de JONG, G., LAMMERS, N., SPRUIT, F. (1986) LC/GC Magazine, p. 1162

57. GROB, Jr., K., STOLL, J.-M. (1986) J. High Res. Chromatogr. **9**: 518

58. GROB, Jr., K., MÜLLER, E., MEIER, W. (1987) J. High Res. Chromatogr. **10**: 416

59. PACCIARELLI, B., MÜLLER, E., SCHNEIDER, R., GROB, Jr., K., STEINER, W., FRÖHLICH, (1988) J. High Res. Chromatogr. **11**: 135

60. GIANESELLO, Y., BOLZANI, L., BRENN, E., GAZZANIGA, A. (1988) J. High Res. Chromatogr. **11**: 99

61. CORTES, H. J., PFEIFFER, C. D., RICHTER, B. E. (1985) J. High Res. Chromatogr. **8**: 469

62. DUQUET, D., DEWAELE, C., VERZELE, M., McKINLEY, S. (1988) in Proc. 9th Int. Symp. on Capillary Chrom., Huethig Verlag, Heidelberg, p. 450

63. KLESPER, E. CORWIN, A. H., TURNER, D. A. (1962) J. Org. Chem. **27**: 700

64. KLESPER, E. (1988) Fresenius Z. Anal. Chem. **330**: 200

65. GERE, D. R., HOUCK, R. K., PACHOLEC, F., ROSSELLI, A. C. P. (1988) Fresenius Z. Anal. Chem. **330**: 222

66. SCHMITZ, F. P. (1988) Fresenius Z. Anal. Chem. **330**: 229

67. ENGELHARDT, H. (1987) in "on-column" (Carlo Erba instruments) 2/87

II   Analysis of Food Ingredients

II.1   Components in Foods with a High
        Carbohydrate Content

II.2   Fats and Compounds in Fat Containing
        Foods

II.3   Aroma Compounds

II.4   Stereodifferentiation of Chiral
        Flavor and Aroma Compounds

II.5   High Molecular Weight Compounds

II.1 Components in Foods With a High Carbohydrate Content

O. Fröhlich

1 Introduction

The capillary column, now available for more than 20 years [1], has not been generally accepted as a universal tool in the analysis of food components. The Association of Official Analytical Chemists (AOAC) has started to suggest the use of capillary columns in their "Official Methods of Analysis" [2] for the determination of natural compounds in 1985.

Nowadays, in the analysis of flavor components, separation is mostly achieved by high resolution gas chromatography (HRGC) using capillary columns [3].

Due to the complexity of natural food material it is necessary to isolate the fraction of interest to obtain a maximum amount of qualitative and quantitative information. A general scheme of sample preparation is outlined in Figure II.1.–1.

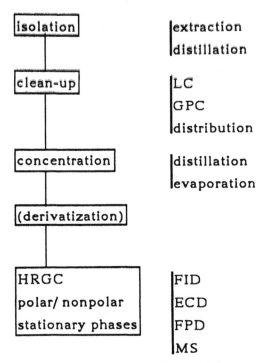

Fig. II.1.–1 General scheme of sample preparation for HRGC.

In the case of foods with a high carbohydrate content the generally necessary sample clean-up can be reduced to a minimum, because carbohydrates are not extractable with organic solvents and not distillable without derivatization. Analyses of components in foods with a high fat content sometimes cause severe problems in sample preparation, which is discussed in chapter II.2.

The following examples give a survey on the use of capillary gas chromatography in the analysis of components in foods with a high carbohydrate content and demonstrate the use of the different methods of sample preparation and detection, shown in Figure II.1.–1. The procedures may easily be modified for the analysis of similiar compounds in other foodstuffs than mentioned.

2 Practical Examples

2.1 Determination of Propionic Acid in Bread

Propionic acid is widely used as a preservative in packed cut bread. The maximum amount legally tolerated is 3.0 g/kg [4]. Isolation is achieved via steam distillation of the ·acidified bread sample [5]. Before concentration of the aqueous distillate is performed, addition of soda lye is necessary to reduce the volatility of propionic acid. The most important steps in sample preparation are shown in Figure II.1.–2.

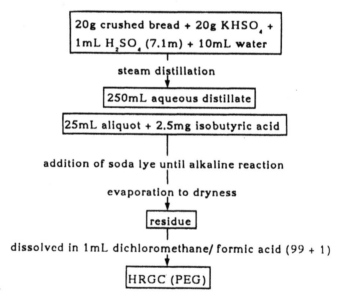

Fig. II.1.–2 Sample preparation scheme for the determination of propionic acid in bread [5].

Separation can be achieved on a capillary column with a polyethyleneglycol (PEG) stationary phase (Carbowax-20 M, FFAP, etc.). The method is reproducible for 0.08 g propionic acid/kg dried bread. The same method is applicable for the determination of propionic
acid in baker's ware [6].

2.2   Determination of Antioxidants in Chewing Gum

In many analytical methods for the determination of 2,6-di-(dimethylethyl)-4-methylphenol
(BHT) and the 2- and 3-dimethylethyl-4-hydroxyanisole isomers (BHA) in fats, oils, soaps,
and cereals, packed column techniques have been used as reported by FRY and WILLIS
[7], ISSHIKI et al. [8], MIN et al. [9], SEDA and TONINELLI [10], WYATT [11] and AOAC Method
20.012 [12].

For the determination of BHA and BHT in chewing gum GREENBERG et al. [13] have introduced a method, which successfully resolves antioxidants from flavor oil volatiles and
gum base components (cf. Fig. II.1.–3).

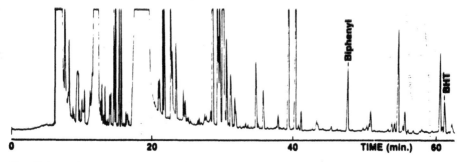

Fig. II.1.–3 Gas chromatogram of peppermint flavored gum extract [13].

For analysis chewing gum is dissolved in toluene as shown in Figure II.1.–4. The gum
base polymer components are precipitated by addition of 2-propanol, containing the
internal standard compound biphenyl. The sample is centrifuged, concentrated by distillation, and further purified by passage through a Waters Sep-Pac Florisil cartridge prior to
injection. A 60 m × 0.25 mm i.d. fused silica capillary column, coated with a thick film (1.0
μm) of DB-1 (J&W Scientific, Folsom, CA, USA) is used for chromatographic separation.
The method reported is reproducible at 4.0 ppm and detects concentrations as small as
2.0 ppm. The linear range occurs between 10 and 150 ppm.

2.3   Determination of Artificial Flavors in Chewing Gum

The use of artificial flavors is restricted to certain foods. Depending on the compound
used concentrations range between 1 and 250 mg/kg [14]. Chewing gum is revealed to
be one of the most trouble-causing foods in this type of analysis.

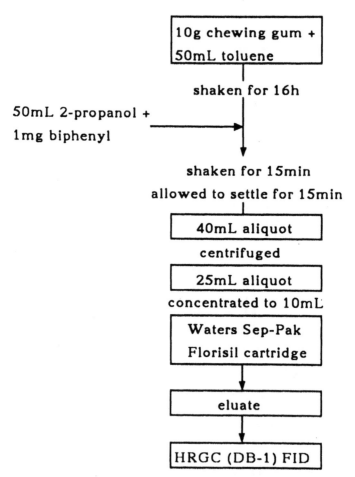

Fig. II.1.–4 Sample preparation scheme for the determination of BHA and BHT in chewing gum [13].

Based on the determination of antioxidants in chewing gum (cf. II.1.2.2) described by GREENBERG et al. [13] a sample preparation procedure for the determination of artificial flavor compounds was developed [15]. In an extract prepared from chewing gum cigarettes (Süsshansa, Trawigo, Würselen) crushed by addition of liquid nitrogen, 98.6 mg/kg ethyl-vanillin were determined with satisfactory recovery values as shown in Table II.1–1. Chromatographic separation could be achieved on a 30 m × 0.25 mm i.d. fused silica capillary column coated with 0.25 μm DB-5 as stationary phase. Split injection (1:10) and flame ionisation detection were applied.

2.4   Determination of Natural Thickening Agents

Natural thickening agents based on polysaccharides are widely used in several industrially produced foods [16, 17]. Their qualitative and quantitative determination can be achieved best by analysis of the sugar monomers, building up to the polysaccharide.

Based on the results reported by Preuss and Thier [18, 19] an official method for the analysis of natural thickening agents and gums in foods has been developed, to be used by food controlling institutes [20].

The method is suitable for agar, carageenan, sodium alginate, locust bean gum, guar gum, gum arabic, pectin, gum tragacanth, carboxymethylcellulose, propyleneglycolalginate, xanthan, gum ghatti, tamarind, and larch arabinogalactan. Fat can be easily removed from these foods by extraction with dioxan. After enzymatic degradation of starch remaining proteins are removed by selective precipitation with sulfosalicyclic acid. Finally, the polysaccharides are precipitated from the aqueous solution by the addition of excess ethanol. After methanolysis the resulting methylglycosides are derivatized by trimethylsilylation. The separation and quantification is performed by gas chromatography using capillary columns with a methylsilicone stationary phase such as SE-30 or DB-1.

Preuss and Thier [18] have analyzed more than 200 samples of commercial thickeners and gums. In most cases the natural sugar composition of the hydrocolloids did not vary more than 10 to 15 %. Figure II.1.–5 shows a gas chromatogram of the derivatized sugar components of all thickeners. Corresponding analytical data are listed in Table II.1.–2.

With the method described it is possible to analyze single thickening agents and gums or mixtures in foods with recoveries ranging from 60 to 85 %. The coefficient of variation ranges between 2 to 8 %. Analytical results for some commercial food samples (Table I.1.–3) demonstrate good agreement with the true values, which were kindly provided by the producers.

Recently Li and Andrews [21] have reported the successful resolution of trimethylsilylated oximes of monosaccharides and uronic acids on a methylsilicone stationary phase, using β-phenyl-D-glucopyranoside as internal standard compound.

The derivatives could easily be prepared by heating the sugars with hydroxylamine hydrochloride dissolved in pyridine and the subsequent addition of hexamethyldisilazane. Starting with dried sugar extract, the entire procedure can be accomplished within 2 hours.

2.5   Determination of Ergot Alkaloids in Cereal Products

The ergot alkaloids have caused severe poisonings of man in the past. The last one in Europe occurred in 1951 in France [22]. Klug et al. [23] have developed a method for the routine analysis of ergometrine, ergometrinine, ergosine, ergosinine, ergotamine, ergota-

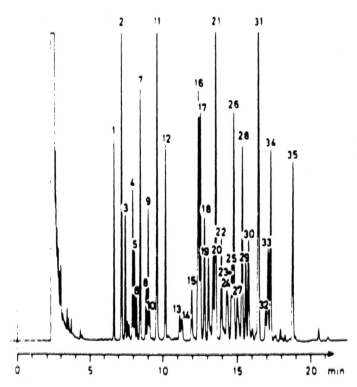

Fig. II.1.–5 Gas chromatogram of all possible sugar components of all thickeners; achieved by menthanolysis of a mixture of gum tragacanth, sodium alginate, xanthan, and gum arabic. – for peak numbers refer to Tab. II.1.–2. [18].

Table II.1.–1 Sample preparation and analysis of artificial flavors in chewing gum [15]

Toluene-extraction (14h); Addition of internal standard compound 2-hydroxyacetophenone in 2-propanol; Centrifugation; Filtration over Sep-Pak (Florisil)	
HRGC-MS:	ethylvanillin
HRGC:	98.6 mg/kg
added amounts:	28 and 61 mg/kg
recovery:	88.2 – 100.7 %

Table II.1.–2 Peak numbers of the methylglycosides of the sugars and uronic acids (internal standards: i-erythrit and sorbit) and relative retention times, based on sorbit (rt = 1.00) [18]

Sugar	Peak No.[1])	Rel. Ret. Time
i-erythrit (i.st.)	1	0.26
arabinose	2, 3, 6	0.29; 0.30; 0.35
rhamnose	4, 5	0.33; 0.34
fucose	5, 7, 9, 10	0.34; 0.37; 0.40; 0.41
guluronic acid	8, 13, 18, 19	0.39; 0.52; 0.61; 0.64
xylose	11, 12	0.43; 0.47
mannuronic acid	14, 17, 24, 25	0.52; 0.60; 0.70; 0.71
glucuronic acid	15, 32, 33	0.57; 0.88; 0.89
galacturonic acid	16, 20, 28, 29	0.59; 0.65; 0.78; 0.79
mannose	21, 23	0.67; 0.70
galactose	22, 26, 27, 30	0.68; 0.74; 0.75; 0.80
glucose	31, 34	0.85; 0.90
sorbit (i.st.)	35	1.00
3.6-anhydrogalactose	36	0.58

Table II.1.–3 Amount of thickeners in food – comparison of determined and true values [19]

Food	Identified Thickeners	Found Value	True g/100 g
ice cream, vanilla	locust bean gum	0.13	0.15
	guar	0.08	0.05
ice cream, lemon	locust bean gum	0.20	0.20
	carboxymethylcellulose	0.13	0.12
sauce powder, vanilla	carrageenan	4.3	5.0
pudding powder, vanilla	not detectable	0	0
tart coating	carrageenan	10.0	11.0
ketchup	guar	0.18	0.15
	locust bean gum	0.12	0.12

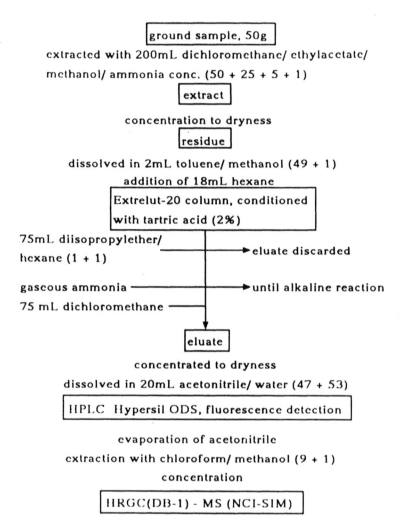

ground sample, 50g

extracted with 200mL dichloromethane/ ethylacetate/ methanol/ ammonia conc. (50 + 25 + 5 + 1)

extract

concentration to dryness

residue

dissolved in 2mL toluene/ methanol (49 + 1)
addition of 18mL hexane

Extrelut-20 column, conditioned with tartric acid (2%)

75mL diisopropylether/ hexane (1 + 1) ──────────► eluate discarded

gaseous ammonia ──────────► until alkaline reaction
75 mL dichloromethane ──────

eluate

concentrated to dryness

dissolved in 20mL acetonitrile/ water (47 + 53)

HPLC Hypersil ODS, fluorescence detection

evaporation of acetonitrile

extraction with chloroform/ methanol (9 + 1)

concentration

HRGC(DB-1) - MS (NCI-SIM)

Fig. II.1.–6 Sample preparation scheme for the determination of ergot alkaloids in food [23].

minine, ergocornine, ergocorninine, α-ergocryptine, α-ergocryptinine, β-ergocryptine, β-ergocryptinine, ergocristine, and ergocristinine. The method consists of extraction, cleaning of the crude extract by a modified form of the Extrelut method, and finally identification and quantitative determination of the alkaloids by high pressure liquid chromatography (HPLC) as shown in Figure II.1.–6. The results are confirmed by HRGC-MS using a methylsilicone stationary phase and single ion monitoring (SIM) after negative chemical

Table II.1.–4 Significant masses of the ergot alkaloids, using NCI-SIM [24]

Alkaloid (Peptide Fragment)	m/z
ergometrine	323/324
ergosine	280/281
ergotamine	314/315
ergocornine	294/295
α-ergocryptine	308/309
β-ergocryptine	308/309
ergocristine	342/343

ionization (NCI). In Table II.1.–4 the significant masses of the ergot alkaloids' peptide fragments are indicated [24]. The acid bond is cut quantitatively using injection port temperatures of about 250 °C.

Market investigations have shown contaminations in ecological as well as in commercial products, so for example many rye products proved to be contaminated. Within the European Economic Community (EEC), a maximum value of 0.05 % ergot is allowed. This corresponds to a total alkaloid content of 1 mg/kg in cereals used for food production. With the method described a minimum amount of 25 to 100 µg/kg can be determined. Using HRGC-NCIMS, a detection limit in the picogram range for semiquantitative determinations can be achieved.

2.6 Determination of Ethyl Carbamate in Alcoholic Beverages

Recently, ethyl carbamate (EC) has attracted attention due to its occurrence in certain alcoholic beverages at levels much higher than would be expected as a result of the fermentation process. This carcinogenic substance [25] has appeared in alcoholic beverages in the past as a result of the use of diethyl pyrocarbonate [26], an antimicrobial agent, which is no longer permitted.

Recently, CHRISTOPH et al. [27] reported on a light-induced reaction which, depending on the concentration of cyanide and benzaldehyde, is responsible for the high amounts of EC up to 10 mg/L [28]. The maximum concentration for alcoholic beverages has been set at 0.8 mg/L in 1986 [29].

For the determination of EC the Extrelut-extraction procedure with butyl carbamate as internal standard compound, published by MILDAU et al. [29] is commonly employed (cf. Fig. II.1.–7). Monitoring the significant fragments of ethyl and butyl carbamate (m/z 44, 62, and 74) with the method described, a detection limit of 10 µg/L can be reached. The results are reproducible at the 20 μg level. In Table II.1.–5 standard variations and coefficients of variation for results, obtained with this method are shown.

Table II.1.–5 Reproducibility of EC determinations in alcoholic beverages at different concentrations

Sample	n	Mean Value (mg EC/l)	Standard Variation at P = 95 % (mg/l)	Coefficient of Variation (%)
Mirabelle brandy	6	3.49	+/– 0.0931	+/– 2.6
Mirabelle brandy	6	0.762	+/– 0.0354	+/– 4.6
Plum brandy	6	0.571	+/– 0.0148	+/– 2.6
Plum brandy	6	0.265	+/– 0.0067	+/– 2.5
White wine	5	0.022	+/– 0.0026	+/– 12

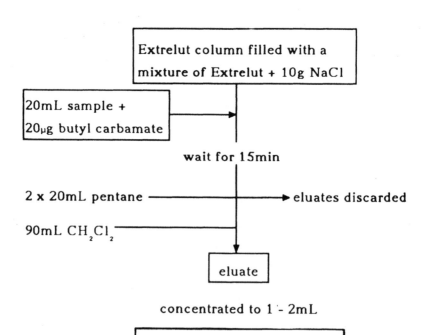

Fig. II.1.–7 Sample preparation scheme for the determination of ethyl carbamate in alcoholic beverages [29].

Another rapid gas chromatographic method for determining EC employs thermal energy analyzer detection in the nitrogen mode [30]. The extraction step used in this study has been adapted from a rapid column elution method for the determination of N-nitrosodimethylamine in malt beverages [31]. Briefly, the beverage is mixed with celite and packed in a column containing a layer of alumina capped with a layer of sodium sulfate. EC is then eluted with methylene chloride, thus combining extraction and clean-up into one step.

Recoveries and standard deviations of EC in wine and whisky fortified at the 20 and 133 µg/kg (ppb) levels averaged to 87.3 +/- 5.3 and 88.7 +/- 3.6 %, respectively. The method has a limit of detection of 1.5 ppb.

2.7 Determination of β-Asarone in Alcoholic Beverages

β-Asarone ((Z)-1,2,4-trimethoxy-5-(1-propenyl) benzene) is a bitter tasting compound in extracts, obtained from *Acorus calamus*, which are used for flavoring alcoholic beverages. Due to its toxicity [32] and mutagenicity [33] a maximum concentration of 1 mg/L is permitted.

For the qualitative and quantitative analysis of β-asarone in alcoholic beverages, we developed a procedure, applying the so-called "solid phase extraction" (SPE) for isolation [34] as outlined in Figure II.1.–8. Quantification is achieved by adding β-asarone to the sample matrix. The recovery values range from 82 to 85 %.

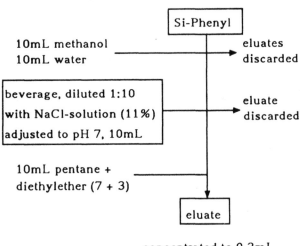

Fig. II.1.–8 Application of SPE for the determination of β-asarone in alcoholic beverages [35].

Using the method described, concentrations up to 3.5 mg/L were determined in commercial beverages exceeding the legal limit in some cases by more than 300 %.

In the recommended method 21/1987, published by the International Organization of the Flavor Industry (I.O.F.I.) for determination of β-asarone by capillary gas chromatography, steam distillation followed by extraction with isooctane is applied for isolation [36]. α-Asarone ((E)–1,2,4–trimethoxy–5–(propenyl) benzene) is suggested as suitable internal standard compound. However, its occurrence in natural extracts as well as its fast isomerization into β-asarone caused by exposure to light and heat [35] has not been taken into account in this procedure.

With the selective isolation by SPE using phenylsubstituted silica gel stationary phases, much better results can be achieved in a shorter time.

2.8 Determination of Halogenated Acids in Beer and Wine Products

Monohalogenated acids are used as free acids or esters for the disinfection of tanks, filters and bottling machines in breweries. If tubes are not carefully flushed with water afterwards, halogenated acids can contaminate the products.

For the determination of monobromoacetic acid a number of methods are known, using microbiological tests, determination of organically bound bromine, or paper- or thin layer chromatography [37, 38]. Recently, a headspace- and a HRGC-MS procedure have been reported [39, 40].

For the simultaneous determination of chloro-, bromo-, and iodoacetic acid the method published by GILSBACH [41] is widely used. The samples are extracted with diethylether which results in recoveries in excess of 70 % for bromoacetic acid and better than 80 % for chloroacetic acid. After derivatization of the concentrated extracts with ethanol and sulfuric acid, separation of the ethyl esters is achieved on a capillary column coated with Carbowax-20 M. Electron capture detection leads to a detection limit of 0.02 mg/L for chloroacetic acid, 0.001 mg/L or below for bromo- and iodoacetic.

2.9 Determination of Diethylene Glycol in Wine

In 1985 the wine market was shocked by findings of the toxic, mutagenic, teratogenic, and carcinogenic compound diethylene glycol (DEG) [42].

A fast and simple method for its qualitative and semiquantitative determination is thin layer chromatography [43]. For concentrations below 1 mg/L capillary gas chromatography is necessarily required. Direct injection of the wine samples [44, 45, 46] leads to contamination of injection system (glass liner, retention gap) and column, resulting in faulty base lines, poor resolution, and production of artefacts [47].

To overcome these problems, sugars and acids can be precipitated as their barium salts [48], as reported by REBELEIN [49] for determination of glycerol and 2,3-butanediol. The wine sample is added to a mixture of sand and barium hydroxide. The mixture is shaken several times. After 5 min acetone is added and the mixture is kept for 5 min at 45 °C. After filtration and evaporation to dryness, the residue is dissolved in a specific volume of ethanol (20 %) to be used for determination on a PEG stationary phase. Quantification is achieved with an external standard.

Another possibility is the silylation of the sample prior to gas chromatographic separation as reported by ROTTSAHL [50]. The wine sample is dried on silica gel at 40 °C in vacuo. The DEG is desorbed with acetone and, after silylation, separated on an OV-101 capillary column. Ethylene glycolmonophenylether is used as internal standard. A detection limit < 10 mg/L with a recovery of 85 % has been reported.

For the determination of trace amounts of DEG an isotope dilution assay can be used. This method was first applied to the determination of glucose in biological material [51].

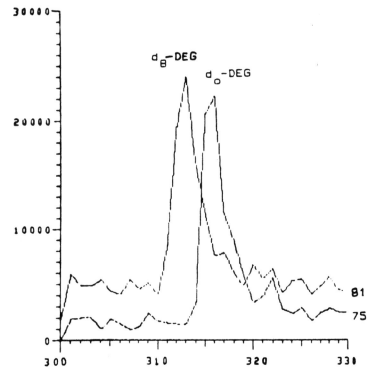

Fig. II.1.–9 Mass fragmentogram of the fragments m/z 75 an 81, recorded from a mixture of d_0-DEG/d_8-DEG (5 ng each) [53].

For the quantification of DEG, d_8-DEG is used as internal standard compound [52]. Monitoring the significant masses m/z 45, 75 and 76 for DEG and m/z 49, 81, and 82 for d_8-DEG a detection limit of 0.5 mg/L can be achieved. Figure II.1.–9 shows the mass fragmentograms of m/z 75 and 81, which are mostly used for quantitative determination. No discrimination of hydrogen rearrangements versus deuterium-transfer could be observed during fragmentation of DEG and d_8-DEG [53]. The linearity of the signal intensity ranges from 1 to 100 ng, i.e. 1 to 100 mg/L when injecting 1 μL.

3 Conclusions

Analytical methods using capillary gas chromatography have been developed to indispensible tools in the analysis of food ingredients. The examples represented here demonstrate their capabilities for determination of components in foods with a high carbohydrate content. The different ways of sample preparation as well as universal or selective detection methods have been discussed.

References

1. JENNINGS, W. (1978) "Gas Chromatography with Glass Capillary Columns", Academic Press, New York.

2. J. Assoc. Off. Anal. (1985) **68**: 1273.

3. JENNINGS, W., TAKEOKA, G. (1984) in "Analysis of volatiles", (Schreier, P., Ed.) W. de Gruyter, Berlin, New York, p. 63.

4. Zusatzstoff-Zulassungs-Verordnung vom 22. 12. 1981 (BGBI. I S. 1633) i.d.F. vom 20. 12. 1984 (BGBI. I S. 1652) Anlage 3 zu § 3.

5. Amtliche Sammlung von Untersuchungsverfahren nach § 35 LMBG L.17.00-14, Beuth, Berlin, Köln.

6. Amtliche Sammlung von Untersuchungsverfahren nach § 35 LMBG L.18.00-11, Beuth, Berlin, Köln.

7. FRY, J. C., WILLIS, N. R. T. (1982) Leatherhead Food R. A. **763**: 1.

8. ISSHIKI, K., TSUMURA, S., WATANABE, T. (1980) Agric. Biol. Chem. **44**: 1601.

9. MIN, D. B., TICKNOR, D., SCHWEIZER, D. (1982) J. Assoc. Off. Anal. Chem. **59**: 378.

10. SEDEA, L., TONINELLE, G. (1981) J. Chromatogr. Sci. **19**: 290.

11. WYATT, D. M. (1981) J. Assoc. Off. Anal. Chem. **58**: 917.

12. AOAC, (1980) "Official Methods of Analysis," 13th ed. Association of Official Analytical Chemists, Washington, D. C.

13. GREENBERG, M. J., HOHOLICK, J., ROBINSON, R., KUBIS, K. GROCE, J., WEBER, L. (1984) J. Food Sci. **49**: 1622.

14. Aromenverordnung vom 22. 12. 1981 (BGBI. I S. 1625, 1676), i.d F. vom 10. 7. 1984 (BGBI. I S. 897) Anlagen 2 und 3 zu § 3.

15. SCHREIER, P., GÖTZ-SCHMID, E.-M., GÜNTHER, C., FRÖHLICH, O. (1987) "Erarbeitung einer Methode zum Nachweis und zur Bestimmung der in Anlage 2 Nr. 1a (zu § 3) der Aromenverordnung aufgeführten und zur Herstellung der in Anlage 3 genannten Lebensmittel begrenzt zugelassenen Aromastoffe". MvP-Hefte 2/1987, Max von Pettenkofer-Institut des Bundesgesundheitsamtes, Berlin.

16. KUHNERT, P. (1980) Gordian **80**: 172.

17. SANDERSON, G. R. (1981) Food Technol. **35/7**: 50.

18. PREUSS, A., THIER, H.-P. (1982) Z. Lebensm. Unters. Forsch. **175**: 93.

19. PREUSS, A., THIER, H.-P. (1983) Z. Lebensm. Unters. Forsch. **176**: 5.

20. Amtliche Sammlung von Untersuchungsverfahren nach § 35 LMBG L.10.00-13, Beuth, Berlin, Köln.

21. LI, B. W., ANDREWS, K. W. (1986) Chromatographia **21**: 596.

22. MÜHLE, E., BREUEL, K. (1977) "Das Mutterkorn" Ziemsen, Wittenberg.

23. KLUG, C., BALTES, W., KRÖNERT, W., WEBER, R. (1988) Z. Lebensm. Unters. Forsch. **186**: 108.

24. KLUG, C. (1986) "Bestimmung von Mutterkornalkaloiden in Lebensmitteln". MvP-Hefte 2/1986. Max von Pettenkofer-Institut des Bundesgesundheitsamtes, Berlin.

25. SCHMAHL, D., PORT, R., WAHRENDORF, J. (1977) Int. J. Cancer **19**: 77.

26. LOFROTH, G., GEJVALL, T. (1971) Science (Washington, D. C.) **174**: 1248.

27. CHRISTOPH, N., SCHMITT, A., HILDENBRAND, K. (1987) Alkohol-Industrie **16**: 269.

28. CHRISTOPH, N., SCHMITT, A., HILDENBRAND, K. (1986) Alkohol-Industrie **15**: 347.

29. MILDAU, G., PREUSS, A., FRANK, W., HEERING, W. (1987) Dtsch. Lebensm.-Rdsch. **83**: 69.

30. CANAS, B. J., HAVERY, D. C., JOE, F. L. (1988) J. Assoc. Off. Anal. Chem. **71**: 509.

31. HOTCHKISS, J. H., HARVEY, D.C., FAZIO, T. (1981) J. Assoc. Off. Anal. Chem. **64**: 929.

32. ABEL, G. (1987) Planta Med. **7**: 251.

33. JIN, Z., QUIN, W., CHEN, X. (1982) Zhejiang Yike Daxue Xuebo **11 (1)**: 1.

34. SCHREIER, P., FRÖHLICH, O. (1987) in "Schnellmethoden zur Beurteilung von Lebensmitteln und ihren Rohstoffen" (Baltes, W., Ed.) Behr's Verlag, Hamburg.

35. LANDER, V. Diss. Universität Würzburg, (in prep.).

36. I.O.F.I. Recommended Method 21/1987 (1988) Z. Lebensm. Unters. Forsch. **186**: 36.

37. MERGENTHALER, E. (1962) Z. Lebensm. Unters. Forsch. **119**: 144.

38. HALLER, H. E. JUNGE, C. (1971) Dtsch. Lebensm.-Rdsch. **67**: 231.

39. CHRISTOPH, N., KREUTZER, P., HILDENBRAND, K. (1985) Weinwirtsch. Techn. **121**: 272.

40. BOECK, K., HARTMANN, A., SPEER, K. (1985) Dtsch. Lebensm.-Rdsch. **81**: 275.

41. GILSBACH, W. (1986) Dtsch. Lebensm.-Rdsch. **82**: 107.

42. ALTMANN, H.-J., GRUNOW, W., KRÖNERT, W., UEHLEKE, H. (1986) Bundesgesundhbl. **29**: 141.

43. LEHMANN, G., GANZ, J. (1985) Z. Lebensm. Unters. Forsch. **181**: 362.

44. FÜHRLING, D., WOLLENBERG, H. (1985) Dtsch. Lebensm.-Rdsch. **81**: 325.

45. LITTMANN, S. (1985) Dtsch. Lebens.-Rsch. **81**: 325.

46. Information ict-Handelsges., Frankfurt.

47. FOSTEL, H. (1985) Ernährung/Nutrition **9**: 783.

48. WAGNER, K., KREUTZER, P. (1985) Weinwirtsch. Techn. **7**: 213.

49. REBELEIN, H. (1957) Z. Lebensm. Unters. Forsch. **105**: 296.

50. ROTTSAHL, H. (1986) Dtsch. Lebensm.-Rdsch. **81**: 148.

51. SWEELEY, C. C., ELLIOT, W. H., FRIES, I., RYHAGE, R. (1966) Anal. Chem. **38**: 1549.

52. RIDDER, M. (1985) Weinwirtch. Techn. **121**: 82.

53. VEITH, H. J., FISCHER, M., HUA, J., ROTH, H. (1986) Dtsch. Lebensm.-Rdsch. **82**: 257.

II.2 Fats and Compounds in Fat Containing Foods

O. Fröhlich

1 Introduction

As a result of many technical improvements and innovations, the reliability, accuracy, and versatility of modern capillary gas chromatography have undergone a revolution in the past decade [1]. Automation in sample application, analytical control, and data handling through microprocessor control systems have become a routine within the basic procedure. Though capillary columns supply higher resolution, higher analysis speed, and higher efficiency [2] a large number of national and international standards for the analysis of food ingredients are still based on packed column separation [3, 4, 5].

In the analysis of fats and fat containing foods the capillary column will overshadow the packed column sooner or later, as the examples in the following chapters show. Even in

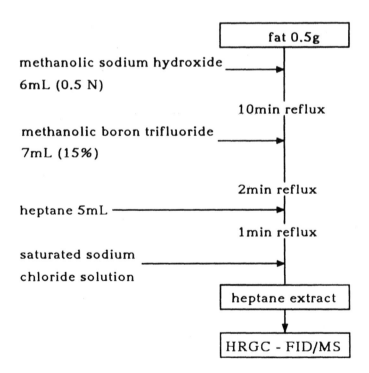

Fig. II.2.–1 Preparation of fatty acid methyl esters using BF₃ [17].

the analysis of light gases, the domain of packed column gas-solid chromatography, support coated open tubular (SCOT) wide bore and capillary columns are available today [6, 7] and have been used successfully [8].

As shown in Figure II.1.–1 extraction with organic solvents is a basic procedure in sample preparation. Due to their lipophilic nature fats are always extracted with the compounds of interest and have to be separated as discussed later.

In the analysis of fats, i.e. the analysis of the fatty acids, building blocks of the triglycerides, derivatization plays an important role, though underivatized separation on new stationary phases has been also achieved [9].

Trimethylsilylated derivatives show excellent gas chromatographic properties [10], but unfortunately, trimethylsilyl esters of acids are rather sensitive to hydrolysis [11]. For this reason analysis of the fatty acid methyl esters (FAME) is preferred, as discussed in the following chapter.

2 Considerations on the Preparation of Fatty Acid Methyl Esters

Many methods for the esterification of fatty acids have been published. They are based on diazomethanolysis [12], acid-catalyzed methanolysis with hydrochloric acid-methanol [13], boron trifluoride-methanol [14, 15] or sulphuric acid-methanol [16]. Also use is made of base-catalyzed transesterification with sodium methoxide-methanol [17–21]. Other techniques involve saponification followed by either acid-catalyzed esterification [17] or boron trifluoride-methanol esterification [21, 22]. Recently, a base-catalyzed transesterification with tetramethylammonium hydroxide has been published [23]. Also, several studies have been published in which different methods are discussed and/or compared [22, 24–28]. It must be concluded that a great many of the published methods have their disadvantages.

The method most commonly employed uses boron trifluoride in methanol as catalyst [15] and is the basis of a number of national and international standards [17, 29, 30]. As shown in Figure II.2.–1, derivatization of fats with neutral pH value can easily be achieved after saponification. It has been proven that acids with low molecular weights are underestimated by this method [31].

At present base-catalyzed methanolysis with sodium methoxide or methanolic potassium hydroxide is widely used in the preparation of FAME from glycerides [17–20]. Figure II.2.–2 shows the sample preparation steps. The acid-catalyzed methylation step is necessary, when appreciable amounts of free fatty acids (FFA) are present, e.g. cheeses with appreciable lipolysis (neutralization value > 2 [32]). Suitable internal standard compounds are the fatty acids C_5, C_{15}, or C_{17}.

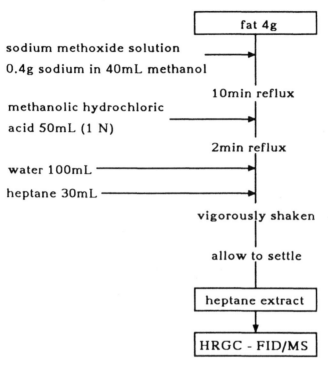

Fig. II.2.–2 Preparation of fatty acid methyl esters using sodium methoxide and acid-catalyzed esterification [17].

Quarternary ammonium hydroxides have been used for the pyrolysis methylation of organic acids [33] and for base-catalyzed transesterification of fats and oils [34, 35]. METCALFE and WANG [23] have proposed a procedure for obtaining the methyl esters of both FFA and glycerides from a sample using tetramethylammonium hydroxide (TMAH) as a catalyst (c.f. Fig. II.2.–3) in a single step and in separate phases. Since TMAH is a strong base, glycerides are transesterified rapidly at room temperature, wheras FFA are converted to TMA soaps, which are transformed to methyl esters in the hot chromatographic injector.

Based on these results analysis of milk fat, fortified with 10 % FFA, has been presented [36]. The analytical results have been in agreement with well-established procedures, such as the KOH-method as shown in Table II.2.–1. Problems may occur with fats containing polyunsaturated fatty acids. The addition of volatile methyl esters (formate or acetate) as well as the use of trimethyl-(α,α,α-trifluoro-m-tolyl)ammonium hydroxide instead of TMAH is recommended.

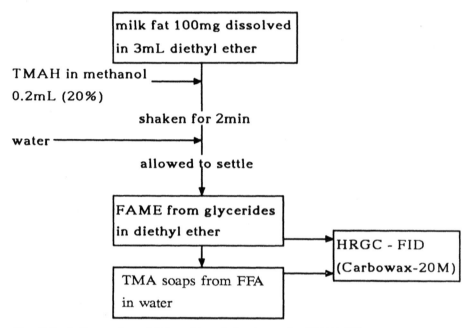

Fig. II.2.–3 Preparation of fatty acid methyl esters using TMAH [23].

Table II.2.–1 Major components (%) of milkfat sample obtained by transesterification in methanol using KOH or TMAH as catalyst [36]

Method	C_4	C_6	C_8	C_{10}	C_{12}	C_{14}	C_{16}	C_{18}	$C_{18:1}$	$C_{18:2}$
KOH x[1])	3.72	2.45	1.44	3.06	4.21	11.82	30.24	9.12	22.31	1.71
s[2])	0.34	0.02	0.16	0.11	0.15	0.30	1.04	0.22	0.36	0.81
TMAH x	3.84	2.69	1.49	3.07	4.13	12.42	30.86	9.88	22.28	1.23
s	0.39	0.39	0.11	0.12	0.27	0.50	0.22	0.38	0.43	0.05

[1]) mean values

[2]) standard deviations

Methylation by treatment with etheral diazomethane solution [13] often leads to artefacts [37]. Diazomethane not only attacks the acidic hydrogens of acids, phenols, or enols as desired, but also adds to the double bonds of α, β-unsaturated esters [38, 39] and reacts with the carbonyl function of α-oxoacids [40]. Figure II.2.–4 shows a gas chromatogram of products formed by the reaction of 2-oxoglutaric acid with diazomethane. The reactions and reaction products have been described [37].

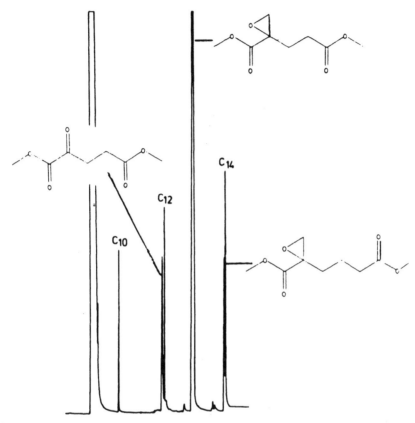

Fig. II.2.–4 Gas chromatogram of products formed by the reaction of 2-oxoglutaric acid with diazomethane [27].

3   Examples for the Analysis of Foods Using Fatty Acid Methyl Esters

3.1   Fatty Acid Composition

In spite of the developments in High Pressure Liquid Chromatography (HPLC), the method of choice today for determining the fatty acid composition of fats and oils is still the gas chromatographic analysis of FAMEs [41].

Triglycerides are base-catalyzed transesterified or methylated after neutralization and saponification [17] as mentioned above. Peak area ratios of specific FAMEs are determined to characterize single vegetable fats or to determine animal fats in vegetable fats by their content of methylbranched FAMEs, not occurring in plants [42].

3.2 Erucic Acid Content

Erucic acid, a characteristic compound in fats of *Brasicacea* species, is determined via quantitative FAME analysis by gas chromatography as described above [17]. A suitable internal standard has to be used, which does not interfere with the erucic acid methyle-ster peak or other components present.

3.3 Butyric Acid Content

The butyric acid content of milkfat, butterfat, or fats containing butterfat although subject to significant natural variation can be assumed to be constant within certain limits for all practical purposes [43]. The European Economic Community (EEC) has adopted a value of 3.6 % for the butyric acid content of milkfat [44].

Butyric acid content is determined via FAMEs' analysis, using valeric acid methyl ester as internal standard [45]. Separation is achieved on a medium polar stationary phase such as cyanopropylsilicone or PEG. With the described method a detection limit of about 0.15 g butyric acid methyl ester/100 g dried bread can be achieved.

3.4 Fatty Acid Composition in Position Two of Triglycerides

Pancreas lipase is used for distinct saponification of the fatty acid esters in positions one and three of triglycerides. The monoglycerides are then separated via liquid chromato-graphy on silica gel and methylated after saponification [46]. The gas chromatographic determination enables a close look at the fatty acid distribution in triglycerides to detect alterations within original fats, e.g. transesterifications.

3.5 (Z)- and (E)-Unsaturated Fatty Acids

After saponification with methanolic NaOH and base-catalyzed methylation with BF_3 a heptane extract is used for gas chromatographic analysis as described earlier. For the chromatographic separation a 60 m × 0.25 mm id SP2340 (Supelco Inc., Bellefonte, PA, USA) fused silica capillary column is necessary [47]. Area percent values for the essential (Z,Z)-9,12-octadecadienoic acid (linoleic acid) and (Z,Z,Z)-9,12,15-octadecatrienoic acid (linolenic acid) are summed to give the total cis-polyunsaturated fatty acid (cis-PUFA) value. Gas chromatographic results have been in good agreement with those obtained by an enzymatic lipoxygenase method [48] at the 31–48 % cis-PUFA levels with a correla-tion coefficient of 0.98. For the method a relative standard deviation of 0.33 % at a 44.4 % cis PUFA level in a margarine sample has been reported [47].

(E)-unsaturated FAMEs can be determined in the same way [49]. By summing the area percent values for all (E)-isomers the total trans-value is determined to give hints at heat abuse or hydrogenation [50].

3.6 Iodine Values

The addition of halogen for analytical purposes was probably first suggested by VON HÜBL [51] a hundred years ago, and has led to the development and application of numerous iodometric methods for the characterization of fats and oils by their degree of unsaturation, each method possessing several specific limitations [52].

Gas chromatographic determination of iodine values (IV) from the fatty acid composition of fats and oils circumvents the problems related to iodometry [53]. A chemical bonded PEG stationary phase capillary column has been used for chromatographic separation of FAMEs. A simple equation which includes double-bond increments of fatty acids, mean molecular weights and molecular weight contributions of each acid has been used for on-line calculation and print out of the results by a computing integrator. Comparative analysis by titration, using the method of WIJS [54] has revealed no significant differences in the results as shown in Table II.2-2.

Table II.2.-2 Comparison of iodometrically and gas chromatographically obtained iodine values [53]

Sample No.	MW	By GC	Wijs	$\triangle$IV
1	276.20	34.12	34.22	− 0.10
2	275.80	34.65	33.97	0.68
3	276.44	38.03	38.02	0.01
4	275.93	39.02	38.58	0.44
5	276.04	40.09	40.33	− 0.24
6	276.01	41.86	41.72	0.14
7	275.94	48.41	47.14	1.27
8	276.82	57.98	59.66	− 1.68
9	275.82	58.16	59.11	− 0.95
10	276.47	59.10	60.49	− 1.39

4   Examples for the Determination of Compounds in Fat Containing Foods

4.1   Determination of Organic Acids of Hens' Eggs

The hygienic conditions of egg products before pasteurization can be evaluated by determination of the contents of succinic acid as a specific indicator for microbial decomposition and β-hydroxybutyric acid indicating fertilized eggs [55, 56]. Both kinds of deterioration are accompanied by an increase in the concentration of lactic acid [57].

The main steps of sample preparation are outlined in Figure II.2.–5. The resulting pattern of organic acids includes lactic acid and citric acid as major components besides pyruvic acid, β-hydroxybutyric acid, fumaric acid, succinic acid, malic acid and pyrrolidone carboxylic acid as shown in Figure II.2.–6.

Recoveries range from 93 to 119 %. Standard deviations for the important acids, succinic acid and β-hydroxybutyric acid, have been found to be 2.6 and 5.6 %, respectively [59].

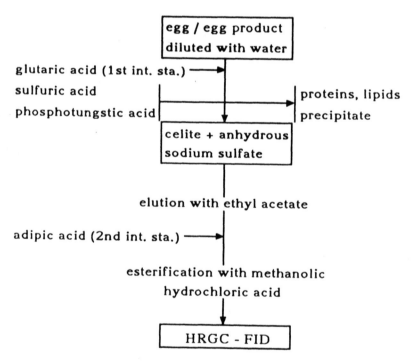

Fig. II.2.–5 Main steps of sample preparation for the determination of organic acids of hens' eggs [58].

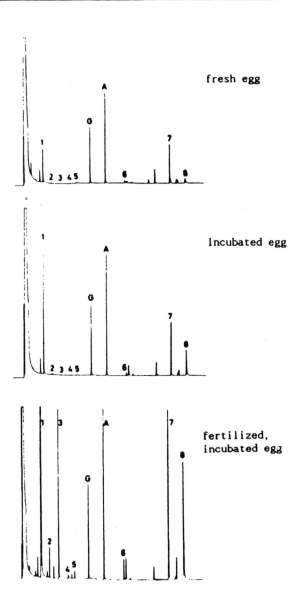

Fig. II.2.–6 Gas chromatogram of methylated organic acids of eggs. **1** lactic acid, **2** pyruvic acid, **3** β-hydroxybutyric acid, **4** fumaric acid, **5** succinic acid, **6** malic acid, **7** citric acid, **8** pyrrolidone carboxylic acid, **A** adipic acid, **G** glutaric acid [57].

4.2 Determination of Artificial Flavors

Generally, artificial flavors can be easily determined in concentrated extracts of foods by HRGC and HRGC-MS [60]. Chewing gum as a food causing trouble during sample preparation has been discussed in chapter II.1. Analysis of fat containing foods, e.g. baker's ware or chocolate via extraction causes problems during gas chromatographic separation due to the amount of fat in the extracts. Based on the successful application of gel permeation chromatography (GPC) for sample clean-up in pesticide analysis [61], SCHREIER et al. have developed a sample preparation procedure to separate fat and flavor compounds [60]. A sample preparation scheme is shown in Figure II.2.–7. The method is able to separate flavor compounds and fat quantitatively as shown in Figure II.2.–8. Gas chromatographic separation of the concentrated extracts has been achieved on a 30 m × 0.25 mm id DB-5 capillary column. Split injection (1:10) and flame ionisation detection have been applied.

Analysis of a chocolate with coconut filling revealed 132 mg/kg of ethylvanillin. Recoveries of samples, fortified with 30, 57, and 86 mg/kg averaged from 91.7 to 108.4 %.

4.3 Detection and Determination of Extraction Oils in Pressed Oils

A procedure for differentiation between virgin (pressed) olive oil and solvent-extracted marc oil has been developed [62]. Wax constituents are separated from the oils by silica gel column chromatography with hexane elution or preparative thin layer chromatography with hexane-diethylether (95:5 v/v).

The wax esters are separated on a short (8 m) capillary column coated with SE-52. Discrimination between marc oils and virgin, lampante, or rectified oils is achieved with the help of a statistical procedure based on C_{34}–C_{46} and amounts of wax esters. Addition of 10 % marc oil to pressed oil is detectable.

4.4 Determination of Cholesterol in Egg Products

Determination of egg content in foods, e.g. baker's ware or egg powder can be achieved via determination of the cholesterol content [63–66]. The use of gas chromatographic determination instead of gravimetric methods leads to better results [67]. In adaption of AOAC method 14141 [68] the cholesterol is determined after acid-catalyzed hydrolysis in the unsaponifiable residue [69]. In Figure II.2.–9 the main steps of sample preparation are outlined. During evaporation to dryness flushing with nitrogen has to be applied to prevent oxidation. Derivatization can be achieved with N-methyl-N-trimethylsilyl trifluoroacetamide (MSTFA) or hexamethyldisilazane/trimethylchlorosilane (HMDS/TMCS) in tetrahydrofuran.

Variation values smaller than 98.27 mg per 100 g +/– 34.73 mg per 100 g in egg powder and 7.41 mg per 100 g +/– 2.62 mg per 100 g in biscuit have been achieved [69]. In Table II.2.–3 a comparison of average cholesterol contents is given, determined gravimetrically as well as gas chromatographically in fresh and dried yolk. The gravimetrically determined values are higher, because accompanied sterols are precipitated together with cholesterol. Using gas chromatography a quantitative separation is possible.

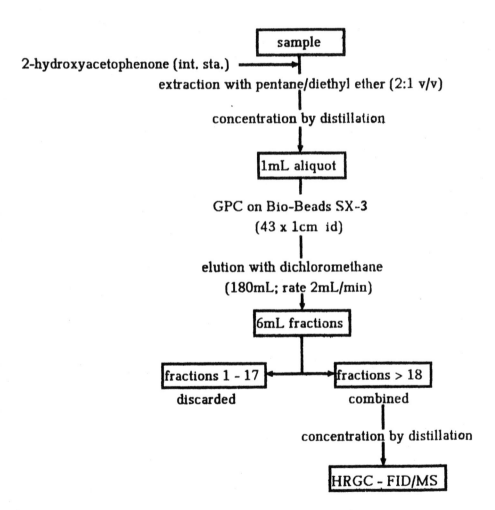

Fig. II.2.–7 Sample preparation scheme for the determination of artificial flavors.

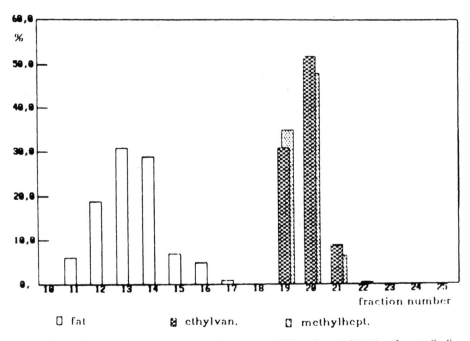

O fat   🅱 ethylvan.   🗋 methylhept.

Fig. II.2.–8 Recoveries of fat, ethylvanillin, and methylheptine carbonate of a synthetic mixture after GPC on Bio-Beads SX-3.

Table II.2.–3 Comparison of cholesterol contents of fresh and dried yolk (% of the dried sample) [70]

Yolk	Method	Sample No.	Mean Value	Range	Standard Deviation	Coefficient of Variation(%)
Fresh	grav.[1]	53	2.56	2.28 – 2.87	0.16	6.4
	GC	53	2.53	2.27 – 2.84	0.17	6.7
Dried	grav	35	2.64	2.26 – 3.05	0.17	6.4
	GC	21	2.61	2.31 – 3.20	0.18	7.0

[1] according to [71]

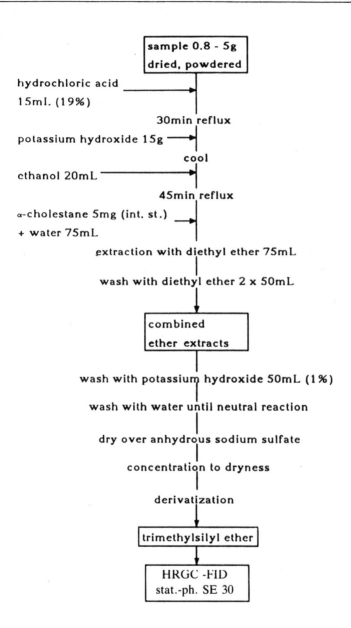

Fig. II.2.–9 Sample preparation scheme for the determination of cholesterol in egg products [67].

4.5 Determination of Sterols and Steryl Esters

In the same way as already mentioned a determination of the whole sterol spectrum is possible [72]. After base-catalyzed saponification and diethyl ether extraction a preparative thin layer chromatographic pre-separation on silica gel can be applied to fractionate the unsaponifiable residue. The sterol fraction is then isolated, extracted, derivatized and separated by HRGC.

The sterols present in the ester form in vegetable oils can be determined separately. After combined LC/TLC the steryl fraction is saponified, derivatized and analyzed by HRGC [73].

4.6 Identification of Brominated Vegetable Oils

Brominated vegetable oils (BVOs) are additives used in citrus-based beverages to disperse the flavoring oils. Though flavored soft drinks are not foods with a high fat content, determination of BVOs is discussed here, because analysis is based on fatty acid methyl ester separation [74].

The beverages are extracted several times with diethyl ether after saturation with sodium chloride. The combined organic extracts are washed with 2 M sodium hydroxide, 2 M hydrochloric acid, and several times with water. The ether phase is dried over anhydrous sodium sulfate and then treated with sulfuric acid methanol for transesterification. Methylheptadecanoate is added as internal standard compound.

Figure II.2.–10 shows a chromatogram of a soft drink analysis. Although many peaks are present, the internal standard, 9,10-dibromostearate methyl ester (DBS) and 9,10,12,13-tetrabromosterate methyl ester (TBS) can be easily detected. The peak area ratio for TBS/DBS for this drink has been 0.87 suggesting the presence of a commercial preparation of brominated partially hydrogenated soybean oil such as Akwilox 133 (Swift Specialty Chemicals, Chicago, IL, USA). No peaks at retention times equivalent to DBS or TBS have been observed in drinks that did not contain BVO [74].

5 Conclusions

The examples for the analysis of fats and compounds in fat containing foods mentioned in this chapter demonstrate that the use of capillary gas chromatography offers definite advantages over existing methods, in terms of resolution, stability, and ease of handling. Even fat containing extracts can be analyzed successfully after separation by gel permeation chromatography, playing an important role in contaminant, pesticide, and flavor analysis.

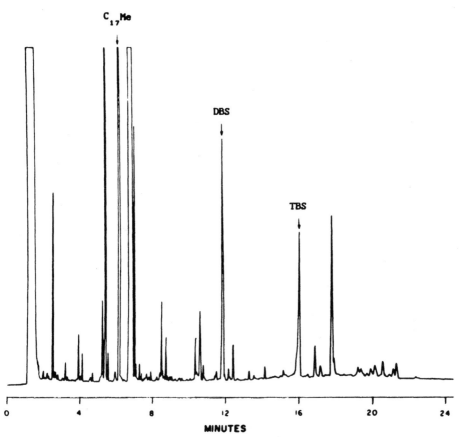

Fig. II.2.–10 Gas chromatogram of an extract of an orange flavored soft drink containing a BVO [74].

References

1. JENNINGS, W. (1985) Proc. 6th Int. Symp. on Capillary Chrom. (Sandra, P., Bertsch, W., Eds.) Huethig Verlag, Heidelberg, p. 1.

2. SCHOMBURG, G. (1987) "Gaschromatography", 2nd ed., VCH, Weinheim.

3. Amtliche Sammlung von Untersuchungsverfahren nach § 35 LMBG, Beuth, Berlin, Köln.

4. AOCS (1980) Official and Tentative Methods of the American Oil Chemists' Society. Champaign, Ill.

5. AOAC (1980) Official Methods of Analysis. Association of Official Analytical Chemists, Washington, D.C.

6. Information (1987) J&W Scientific, Folsom, CA, USA, 11/1987.

7. Information (1987) Chrompack International, Middelburg, The Netherlands.

8. RAUHUT, D. (1987) in 2. Praktina – Europ. Symp. Prakt. Inst. Anal. (David, W., Günther, W., Schlegelmilch, F., Eds.) Hoppenstedt, Darmstadt, Brüssel, Haarlem, Zürich, p. 379.

9. GEERAERT, E., SANDRA, P. (1985) Proc. 6th Int. Symp. on Capillary Chrom. (Sandra, P., Bertsch, W., Eds.) Huethig Verlag, Heidelberg, p. 174.

10. SWEELEY, C. C., BENTLEY, R., MAKITA, M., WELLS, W. W. (1963) J. Amer. Chem. Soc. **85**: 2497.

11. LANGENBECK, U., SEEGMILLER, J. E. (1973) J. Chromatogr. **80**: 81.

12. SCHLENK, H., GELLERMANN, J. L. (1960 Anal. Chem. **32**: 1412.

13. STOFFEL, W., CHU, F., AHRENS, E. H. (1959) Anal Chem. **31**: 307.

14. MORRISON, W. R., SMITH, L. M. (1964) J. Lipid Res. **5**: 600.

15. METCALFE, L. D., SCHMITZ, A. A., PELKA, J. R. (1966) Anal. Chem. **38**: 514.

16. AOAC (1965) Official Methods of Analysis, Washington D.C., 10th ed., Sect. 266.052.

17. Amtliche Sammlung von Untersuchungsverfahren nach § 35 LMBG L.23.04-1, Beuth, Berlin, Köln.

18. CHRISTOPHERSON, S. W., GLASS, R. L. (1969) J. Dairy Sci. **52**: 1289.

19. BADINGS. H. T., DE JONG, C. (1983) J. Chromatogr. **279**: 493.

20. CHRISTHIE, W. W. (1982) Lipid Analysis. 2nd ed. Pergamon Press, New York.

21. BANNON, C. D., BREEN, G. J., CRASKE, J. D., HAI, N. T., HARPER, N. L., O'ROURKE, K. L. (1982) J. Chrom. **247**: 71.

22. VAN WIJNGAARDEN, D. (1967) Anal. Chem. **39**: 848.

23. METCALFE, L. D., WANG, C. N. (1981) J. Chromatogr. Sci. **19**: 530.

24. VORBECK, M. L., MATTIK, L. R., LEE, F. A., PEDERSON, C. S. (1961) Anal. Chem. **33**: 1512.

25. DE MAN, J. M. (1976) Lab. Pract. **16**: 150.

26. IVERSON, J. L., SHEPARD, A. J. (1977) J. Assoc. Off. Anal. Chem. **60**: 284.

27. FIRESTONE, D., HOROWITZ, W. (1979) J. Assoc. Off. Anal. Chem. **62**: 709.

28. SCHUCHMAN, H. (1975) Kontakte (Merck) No 2, 34.

29. AOCS (1980) Official and Tentative Methods of the American Oil Chemists' Society. Champaign, III, Methods Ce 2–66.

30. AOAC (1980) Official Methods of Analysis. Association of Official Analytical Chemists, Washington, D.C. 28.053–28.056.

31. BANNON, C. D., CRASKE, J. D., HAI, N. T., HARPER, N. L., O'ROURKE, K. L. (1982) J. Chromatogr. **247**: 63.

32. Schweizerisches Lebensmittelbuch (1967) 5th ed., vol. 2, chapt. 1/13, Eidg. Drucksachen- und Materialzentrale, Bern, 16.

33. PRELOG, V., PIANTANIDA, M. (1936) Z. Physiol. Chem. **244**: 56.

34. CREARY, D. K., KOSSA, W. C., RAMACHANDRAN, S., KURTZ, R. R. (1978) J. Chromatogr. Sci. **16**: 329.

35. BUTTE, W. (1983) J. Chromatogr.. **261**: 142.

36. MARTINEZ-CASTRO, I., ALONSO, L., JUÁREZ, M. (1986) Chromatographia **21**: 37.

37. BAUER, S., NEUPERT, M., SPITELLER, G. (1984) J. Chromatogr. **309**: 243.

38. VON AUWERS, K., KÖNIG, F. (1932) Anal. Chem. **496**: 252.

39. HUISGEN, R., EBERHARD, P. (1971) Tetrahedron Lett. **45**: 4343.

40. SIMMONDS, P. G., PETTITT, B. C., ZLATKIS, A. (1967) Anal. Chem. **39**: 163.

41. LERCKER, G. (1986) Riv. Ital. Sostanze Grasse **63**: 331.

42. BELITZ, H. D., GROSCH, W. (1982) Lehrbuch der Lebensmittelchemie. Springer, Berlin, Heidelberg, New York.

43. HADORN, H., ZÜRCHER, K. (1970) Dtsch. Lebensm.-Rdsch. **66**: 77.

44. TÖPEL, A. (1981) "Chemie und Physik der Milch", VEB Fachbuchverlag, Leipzig.

45. Amtliche Sammlung von Untersuchungsverfahren nach § 35 LMBG L.17.00-12, Beuth, Berlin, Köln.

46. Metodo NGD C 46-1986 (1986) Riv. Ital. Sostanze Grasse **63**: 404.

47. ATHNASIOS, A. K., HEALY, E. J., GROSS, A. F., TEMPLEMAN, G. J. (1986) J. Assoc. Off. Anal. Chem. **69**: 65.

48. MADISON, B. L., HUGHES, W. J. (1983) J. Assoc. Off. Anal. Chem. **66**: 81.

49. GILDENBERG, L., FIRESTONE, D. (1985) J. Assoc. Off. Anal. Chem. **68**: 46.

50. Metodo NGD C 74-1986 (1986) Riv. Ital. Sostanze Grasse **63**: 301.

51. V. HÜBL, A. (1984) J. Soc. Chem. Ind., London, **3**: 641.

52. POKORNY, J. (1972) Chem. Listy **66**: 22.

53. KOLAROVIC, L., TRAITLER, H., DUCRET, P. (1984) J. Chromatogr. **314**: 233.

54. BOEKENOOGEN, H. A. (Ed.) (1964) "Analysis and Characterization of Oils, Fats and Fat Products", Vol. 1, Interscience, London, 32.

55. SALWIN, H., BOND, J. F. (1969) J. Assoc. Off. Anal. Chem. **52**: 41.

56. DAENENS, P., LARUELLE, L. (1976) J. Assoc. Off. Anal. Chem. **59**: 613.

57. LITTMANN, S., ACKER, L., SCHULTE, E. (1982) Z. Lebensm. Unters. Forsch. **175**: 106.

58. LITTMANN, S. (1985) Dtsch. Lebensm.-Rdsch. **81**: 345.

59. LITTMANN, S., SCHULTE, E., ACKER, L. (1982) Z. Lebensm. Unters. Forsch. **175**: 101.

60. SCHREIER, P., GÖTZ-SCHMIDT, E.-M., GÜNTHER, C., FRÖHLICH, O. (1987) "Erarbeitung einer Methode zum Nachweis und zur Bestimmung der in Anlage 2 Nr. 1a (zu § 3) der Aromenverordnung aufgeführten und zur Herstellung der in Anlage 3 genannten Lebensmittel begrenzt zugelassenen Aromastoffe". MvP-Hefte 2/1987, Max von Pettenkofer-Institut des Bundesgesundheitsamtes, Berlin.

61. THIER, H. P., FREHSE, H. (1986) "Rückstandsanalytik von Pflanzenschutzmitteln", Thieme, Stuttgart.

62. MARIANI, C., FEDELI, E. (1986) Riv. Ital. Sostanze Grasse **63**: 3.

63. ACKER, L. (1967) in "Handbuch der Lebensmittelchemie" (Acker L., Bergner, K.-G., Diemair, W., Heimann, W., Kiermeier, F., Schormüller, J., Souci, S. W., Eds.) vol. V/1, Springer Verlag, Berlin, Heidelberg, New York, p. 471.

64. ACKER, L., GREVE, H. (1964) Z. Lebensm. Unters. Forsch. **124**: 259.

65. HADORN, H., JUNGKUNZ, R. (1952) Mitt. Lebensmittelunters. Hyg. **43**: 1.

66. Mitt. Bundesgesundheitsamt (1962) Bundesgesundh.-Bl. **14**: 223.

67. DRESSELHAUS, M. (1974) "Die gaschromatographische Bestimmung des Cholesterin in Lebensmitteln. – Ein selektives Verfahren zur Ermittlung des Eigehaltes". Diss., Univ. Münster.

68. AOAC (1970) Official Methods of Analysis. Association of Official Analytical Chemists, Washington, D.C., 14141.

69. Amtliche Sammlung von Untersuchungsverfahren nach § 35 LMBG L.05.02-2, L.18.00-10, and L.22.02/04-1 Beuth, Berlin, Köln.

70. DRESSELHAUS, M., ACKER, L. (1974) Lebensmittelchem. gerichtl. Chem. **28**: 355.

71. ACKER, L., DIEMAIR, W. (1957) Z. Lebensm. Unters. Forsch. **105**: 437.

72. DIMITRIOS, B., JOANNA, V. (1986) Lebensm.-Wissen.-Technol. **19**: 156.

73. Metodo NGD C 71-1986 (1986) Riv. Ital. Sostanze Grasse **63**: 304.

74. CHADHA, R. K., LAWRENCE, J. F., CONACHER, H. B. S. (1986) J. Chromatogr. **356**: 441.

II.3 Aroma Compounds

G. Takeoka

1 Introduction

Aroma compounds are an important class of chemicals which largely determine the flavor of foods and beverages. Their analysis has been the subject of a tremendous amount of research effort. Due to the extensive amount of literature available on the topic no attempt will be made to cover all areas comprehensively. Instead, this review is restricted to the chromatographic separation of flavor volatiles.

Aroma chemicals are a structurally diverse group of products which exhibit wide variability in their stability and chromatographic behavoir. Their occurrence as minor constituents in complex matrices has provided a difficult challenge to the flavor chemist. The analysis of flavor chemicals was transformed by the development of gas chromatography. Packed columns were used initially though lacking the resolving power necessary for the separation of these highly complex mixtures. The introduction of capillary columns by Golay in 1957 provided a breakthrough in the field. Our analytical capabilities were further enhanced through the development of fused silica capillary columns with bonded stationary phases and new sample introduction techniques (especially on-column injection). Due to their ease of handling, superior efficiency and higher inertness, fused silica capillary columns are rapidly replacing packed columns in flavor analysis. In efforts to further increase resolution multidimensional GC has been applied in flavor research. The task of compound identification has been greatly facilitated by the use of combined gas chromatography-mass spectrometry (GC-MS). In most cases the combination of mass spectral data and retention indices is sufficient to provide positive identification. However, other spectroscopic techniques such as NMR, IR and UV which require larger sample amounts, may be necessary to confirm identity. Automated micropreparative systems utilizing capillary columns are useful in these cases and their use has been demonstrated.

2 GC System Considerations

The use of clean and inert analytical systems is a prerequisite to obtaining accurate qualitative and quantitative results. Due to the susceptibility of various flavor compounds to degradation and adsorption the use of glass-lined injectors and glass columns, especially fused silica capillary columns is strongly recommended. The repeated injection of complex flavor samples can lead to the accumulation of non-volatile residues in the injector and column inlet. These residues can catalyze the decomposition of subsequent samples injected [1]. Adsorption effects in a dirty injector can lead to tailing peaks and reduced system efficiency [2] as shown in Figure II.3.–1 (A). Increasing the flow rate fourfold during

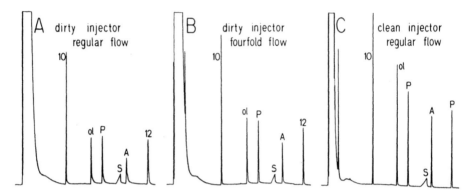

Fig. II.3.–1 Influence of injector cleanliness and gas flow rate on peak symmetry. Compounds were chromatographed on a 20 m × 0.31 mm i.d. SE-52 (d$_f$ – 0.14 µm) column. Regular flow was 2.4 mL/min of hydrogen carrier gas. For B the inlet pressure was increased fourfold without changing the split ratio (1:30). The flow was increased only for injection then reduced to regular flow. Chromatographic conditions were adjusted according to standard Grob test [6]. **10**: decane, **ol**: 1-octanol, **p**: 2,6-dimethylphenol, **2**: 2-ethylhexanoic acid, **A**: 2,6-dimethylaniline, **12**: dodecane [2].

injection (B) improves the peak shapes due to reduced residence time of the sample in the injector. However, the best results are obtained with a clean injector (C). Injector inserts should be checked and cleaned regularly. The type of samples injected and their method of sample preparation influence the frequency of insert cleaning.

Vaporizing injectors subject the sample to thermal shock which may cause decomposition of the more labile compounds. The decomposition of secondary and tertiary bromides in vaporizing injectors has been observed by CARDOSO and ALFONSO [3]. The metal surfaces of a syringe needle can also exert catalytic effects. In their analysis of mustard oil constituents, GROB and GROB, JR, [4] found that degradation products were observed with hot splitless injection. The results were similar when nickel was substituted for stainless steel as the needle material; platinum needles caused even higher levels of degradation. Sample decomposition in a vaporizing injector is illustrated in Figure II.3.–2. No degradation products were observed with on-column injection. Using a modified syringe equipped with a fused silica needle ROERAADE et al. [5] compared its inertness with respect to a conventional syringe with a stainless steel needle. The thermally labile compound, dibenzothiophene-5-oxide, was used to test the influence of syringe needle material in hot splitless injection (250 °C). Far less degradation was observed with the fused silica needle confirming the catalytic activity of the metal surfaces. For samples containing thermally labile solutes and/or high boiling solutes the use of on-column injection is pre-

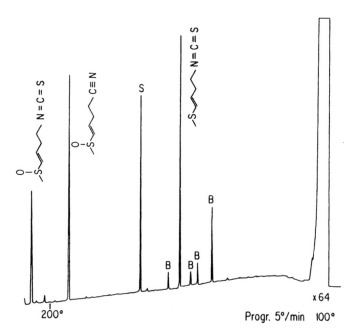

Fig. II.3.–2 Degradation of labile mustard oil constituents with splitless injection. Column
– 15 m × 0.3 mm i.d. Pluronic 64 (d$_f$ – 0.08 µm) on acidic support. Detector:
NPD, N-mode. **B** denotes breakdown products; **S** represents the internal
standard (1-cyanopentadecane, 5 ng/µl). No breakdown products were
observed with on-column injection [4].

ferred. Though the potential of on-column injection was realized in the early days of open-
tubular column chromatography its development as a viable technique occurred much
later with the pioneering work done mainly by the GROBS [6, 7] and SCHOMBURG and co-
workers [8]. A schematic of a GROB on-column injector is illustrated in Figure II.3.–3. The
syringe needle (preferably fused silica tubing) is guided through the conical aperture into
the 0.3 mm channel. The outer diameter of the syringe needle must closely match the
inner diameter of the needle channel in order that no measurable pressure drop occurs at
the column inlet when the stop valve is opened. The syringe needle is guided into the
capillary column and the sample is injected. The syringe needle is then withdrawn to a
point just above the stop valve. After closing the valve the syringe can be fully retracted.
To insure sample transfer from the syringe to the column in the liquid state, cool air is
passed through the bottom part of the injector. The steel beaker serves to isolate the
injector from the warm environment at the oven top. On-column injection offers a number
of advantages over vaporizing injection. First, since the sample is deposited directly on

the column flash vaporization is eliminated, thereby minimizing decomposition due to thermal effects. Second, with the substitution of fused silica needles on syringes the possible catalytic effects of metal surfaces is eliminated. Third, the sample transfer from syringe to column is typically performed in the liquid state eliminating sample discrimination due to syringe effects [9, 10]. Finally, the technique yields highly reproducible results. However, the technique is not suitable for "dirty" samples. With vaporizing injectors the nonvolatile residues are largely retained in the injector liner. In contrast, with on-column injection the direct deposition of sample onto the column presents the more serious problem of accumulation of non-volatile residues in the column inlet. The use of immobilized phase capillary columns permits rinsing the column with solvent to remove the soluble residues. One note of caution is in order. The exposure to solvent causes the stationary phase film to swell; at temperatures above the solvent boiling point violent solvent evaporation can rupture the film [11].The column should be purged with an inert gas at a tem-

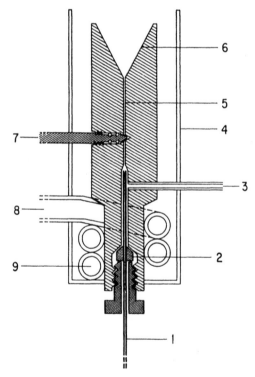

Fig. II.3.–3 Schematic diagram of an on-column injector. **1** Glass capillary column; **2** graphite fitting; **3** carrier gas inlet; **4** steel beaker; **5** 0.3 mm channel; **6** conical aperture; **7** stop valve; **8** coiled copper tubing, cold air in; **9** cold air out [4].

perature below the solvent boiling point for several hours (overnight is convenient) to remove the bulk of the solvent. The temperature can then be raised slowly to remove the remaining solvent. Alternatively, the inlet and outlet ends of the column can be reversed or short sections on the inlet end can be removed to restore column efficiency [12, 13]. The recommended solution is the use of a "retention gap" as suggested by GROB [14]. Here a short section of deactivated, uncoated column tubing (allowing about 30 cm per 1 µl of solvent injected) is attached to the front of the analytical column. The uncoated inlet serves as a pre-column or guard column. Non-volatile residues retained in the inlet are easily eliminated by removing an appropriate length of the uncoated inlet. The length of the analytical column is unchanged, so its efficiency is not affected. If samples containing compounds deleterious to the stationary phase are injected, they do not contact the stationary phase and the resultant contributions to bleeding and peak distortion are avoided. The use of a "retention gap" may also reduce or eliminate the problem of "band broadening in space" associated with splitless and on-column injections at oven temperatures below the boiling point of the sample solvent [14]. The problem is caused by spreading of the sample constituents in the column inlet due to the flow of liquid solvent. The uneven distribution of solute constituents in this flooded zone leads to broadened or split peaks. The retention gap has negligible retention compared to the analytical column. The rapidly moving solute bands are compressed when the more retentive stationary phase film of the separation column is encountered. Retention gaps are important analytical tools. Long retention gaps are useful for on-column injection of large sample volumes. This has obvious benefits in sample analysis. Extracts are frequently concentrated to small volumes to increase the sensitivity. Such concentration efforts can engender solute losses [15]. The injection of large sample volumes simplifies sample preparation and may help to avoid solute losses encountered with conventional sample concentration techniques. The use of long retention gaps has also made the coupling of HPLC and capillary GC feasible [16]. Retention gaps are readily connected to the analytical column through the use of press fit connectors [17]. These glass fittings provide the ideal connection between fused silica capillaries forming a strong, tight and inert seal (see also Chapter I.2., Fig. I.2.–2).

In many applications it is desirable to split the column effluent to allow simultaneous sniffing and selective detection of the separated constituents. The desired features of an effluent splitter are inertness, low dead volume, low thermal mass and the ability to maintain leak-free connections at high temperatures and with repeated thermal cycling. BRETSCHNEIDER and WERKHOFF [18] recently described an all glass splitter for use in capillary gas chromatography. A schematic diagram of the splitting device is shown in Figure II.3.–4. The heart of the splitter is the glass cap tube which is held in place by a graphite ferrule. The fused silica capillaries are inserted into the metal capillaries and terminate in the glass cap. Each fused silica capillary is held in place by its own graphite ferrule. The splitter has been demonstrated to pass large amounts of solvent vapors (on-column injection) with no evidence of peak tailing. The application of the splitter in the simultaneous

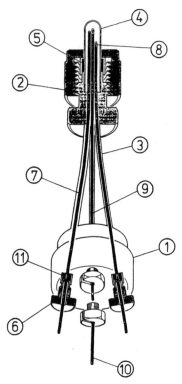

Fig. II.3.–4 Schematic diagram of a "glass cap cross" splitter. Component identities: **(1)** and **(2)** stainless steel body; **(3)** metal capillary; **(4)** glass cap tube; **(5)** graphite ferrule; **(6)** stainless steel nut; **(7)** fused silica capillary column; **(8)** make-up gas line; **(9)** and **(10)** fused silica connecting arms; **(11)** graphite ferrule [18].

detection of meat flavor constituents is illustrated in Figure II.3.–5. This author has used the commercially available "Y" press-fit connectors as effluent splitters and has obtained excellent results. Along with the desirable properties of the above splitter this device is even simpler and easier to use making it the design of choice at present.

Programmed temperature vaporizing (PTV) injection was introduced by VOGT et al. [19] and later refined by POY et al. [20] and SCHOMBURG et al. [21]. The technique involves sample introduction into a cool inlet liner followed by rapid heating after the sample has exited the syringe needle (Fig. II.3.–6). The vaporized sample is then transferred into the column either in the split or splitless mode. The sample transfer under cool conditions overcomes the problems of sample fractionation of higher boiling solutes and catalytic

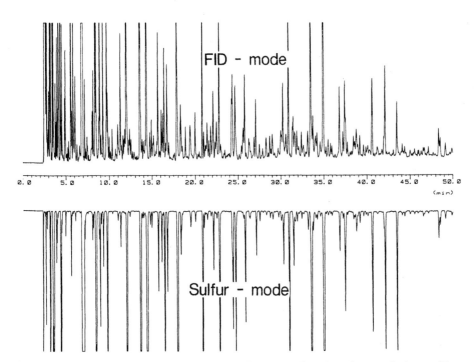

Fig. II.3.–5 Simultaneous detection of meat flavor constituents using an all-glass splitter. Column: 60 m × 0.32 mm i.d. DB-1 (d_f – 0.25 µm). The column temperature was programmed from 60 °C to 230 °C at 3 °C/min. On-column injection was used [18].

degradation of labile solutes due to hot syringe needles [22, 4]. PTV injector volumes are much smaller than conventional vaporizing injectors used for splitless injection. Small internal diameter glass liners (about 0.8–1.1 mm) are required for rapid heating. The small internal volume of the glass liner also allows more efficient sample transfer from the vaporizing chamber to the column. GROB et al. [23] has demonstrated that matrix effects are much smaller in PTV splitless injection than conventional splitless injection; hence samples containing non-volatile constituents are analyzed more accurately and reproducibly using PTV splitless injection than conventional splitless injection. NITZ et al. [24] have utilized a modified PTV inlet to analyze various headspace samples (Fig. II.3.–7). Headspace volatiles were desorbed from a Tenax trap to a liquid nitrogen cooled PTV inlet packed with 5 % OV-101 on Chromosorb W. During this transfer step the split vent is opened to allow a high flow rate for desorption of volatiles from the Tenax trap. WERKHOFF and BRETSCHNEIDER [25] have shown that a fast desorption gas flow rate is necessary for efficient recovery of volatiles from Tenax traps.

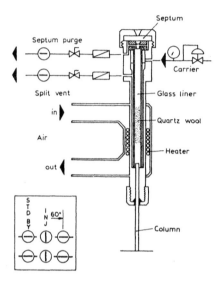

Fig. II.3.–6 Cross-sectional view of a programmed temperature vaporizing (PTV) injector. The shut-off valves for the septum purge and split vent are closed for injection. Following injection the inlet is rapidly heated; after a delay period the shut-off valves are opened [20].

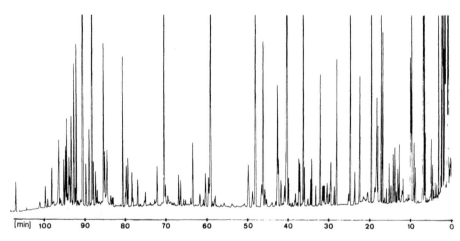

Fig. II.3.–7 Chromatogram of beer wort headspace volatiles analyzed on a 30 m × 0.32 mm i.d. SE-30 quartz capillary column. The volatiles were adsorbed on a Tenax trap. The volatiles were thermally desorbed and cryofocused on a modified PTV inlet. The sample was then transferred to the analytical column [24].

A variety of materials such as plastic, stainless steel, nickel, and other metals have been used in column fabrication. While metal columns are suitable for hydrocarbon analysis their surface activity may cause problems in other applications. BERTSCH et al. [26] described the preparation of nickel columns. The deactivation of nickel capillary columns has been detailed by PRETORIUS et al. [27]. The authors first deposited a layer of elemental silicon on the column surface and completed the deactivation by treatment with octamethylcyclotetrasiloxane (D4). The deactivated nickel tubing displayed good inertness. In spite of their activity toward polar solutes stainless steel columns have been far more popular than nickel columns. It has been reported that ethyl lactate, 2,5-dimethyl-4-hydroxy-3(2H)-furanone, 5-hydroxymethylfurfural and p-allylphenol (Fig. II.3.−8) fail to elute from stainless steel open tubular columns [28]. Similarly, BUTTERY et al. [29] observed the loss of isothiocyanates and other sulfur compounds on stainless steel capillary columns. In their analysis of a lemon oil sample on a stainless steel capillary column AVERILL and MARC [30] noted a baseline rise between the neral and geranial peaks. The rise was probably due to a partial isomerization or decomposition caused by the active metal surface as no baseline change was observed when the sample was run on a glass capillary column.

ethyl lactate   2,5-dimethyl-4-hydroxy-3(2H)-furanone

5-hydroxymethylfurfural

p-allylphenol

Fig. II.3.−8 Solutes which failed to elute from stainless steel open tubular columns [28].

Due to its greater inertness glass is a better choice for column material than metal. Though glass is generally considered to be inert with respect to adsorption and catalytic behavior it can exhibit activity in chromatographic applications. Column wall activity is influenced by the silica surface structure and by metal oxides and metal ions present on the glass surface. These impurities can act as Lewis acid sites and interact with electron donating molecules. The metal oxides are normally removed by acid leaching. In contrast, fused silica has a much lower metal oxide content (< 1 ppm) and a high tensile strength which makes it the column material of choice at present. BLOMBERG and co-workers [31] have noted that silicone phases coated on fused silica have a lower bleed rate than when coated on carefully leached soda glass. The flexibility and inherent straightness of the fused silica capillary column allows its optimum placement in the chromatographic system leading to maximum efficiency and inertness [32]. An excellent review of fused silica capillary column technology recently appeared [33]. Fused silica containing the minimal amount of metal ions and combined water is made from silicone dioxide produced by oxidizing pure silicon tetrachloride in an oxygen electric plasma torch [34].

Glass, however, is still popular in laboratories which prepare their own columns. Glass columns can be drawn on less sophisticated equipment than is required for fused silica and hence their cost is significantly lower in comparison. Additionally, in high temperature applications glass may be the favored material over fused silica. The limiting factor with fused silica is the polyimide outer coating which protects the capillary from corrosion and provides mechanical strength. The polyimide polymer is stable up to about 350 °C. If used above this temperature the polymer slowly begins to thermally degrade. Glass, on the other hand, is limited by the thermal stability of the stationary phase. TRESTIANU and co-workers [35] used short borosilicate glass columns coated with methylpolysiloxane to separate and elute hydrocarbons in the n-C130 to C140 range by using temperature programming up to 430 °C.

By replacing the polyimide outer coating with aluminium LIPSKY and DUFFY [36, 37] produced flexible fused silica capillary tubing able to withstand temperatures up to 500 °C. The aluminium clad columns were coated with methylpolysiloxane and could be operated at 400–425 °C isothermally and 425–440 °C in the programmed mode. This technology extends the analysis of substances with molecular weights up to 1000–1200 and/or boiling points up to 650–750 °C and higher.

3   Stationary Phase Considerations

An important factor in solving a particular application is the choice of the stationary phase. While the use of capillary columns has greatly facilitated the separation of complex flavors, there exist many separation problems that cannot be solved merely by a high number of theoretical plates. The selectivity of the stationary phase has a dramatic effect on the resolution of constituents. With over 200 stationary phases available the analyst may be confused on how to select a phase for a specific separation problem. Many of the available liquid phases were not designed for use in GC; they often possess undesirable chromatographic properties such as wide molecular weight distribution, high batch-to-batch variability and low thermal stability. Efforts to reduce the number of stationary phases have been made. STARK et al. [38] proposed that the following phases would provide a large range of selectivity in capillary GC: (1) methyl-silicone, (2) 50−70 % phenyl methyl-silicone, (3) cyanopropyl methylsilicone of medium (25−50 %) and high (70−90 %) cyanopropyl content, (4) trifluoropropyl methylsilicone and (5) polyethylene glycol.

In the separation of flavor mixtures it is often desirable to employ several phases with different selectivities. To aid in the identification of unknown constituents the analysis on a polar phase such as Carbowax 20M is frequently combined with the analysis on a non-polar phase such as OV-101, SE-30 or DB-1 (Fig. II.3.−9). The determination of retention indices on two different phases can aid in the characterization of unknowns since it is less likely that two constituents will have identical indices on both phases.

The most widely used retention index system is that proposed by KOVATS [39]. The index expresses the retention behavior of a compound as equivalent to that of a hypothetical n-paraffin hydrocarbon. The index is influenced by a number of factors such as film thickness, column temperature, type of cross-linking agent used (for the preparation of bonded phases), column material and age. Using glass capillary columns coated with Carbowax 20M, SHIBAMOTO et al. [40] reported that the Kovats indices of various alcohols, aldehydes and ethyl esters increased as the film thickness increased. They suggested the use of ethyl esters instead of hydrocarbons when determining the retention indices of polar constituents on Carbowax 20M. The influence of column temperature on retention indices was recently discussed by GROB and GROB [41]. Since stationary phase polarity is temperature dependent, changes in solute elution order can occur at different temperatures. This phenomenon can be exploited to resolve overlapping solutes. The magnitude of retention index shift with temperature ($\Delta I/10°$) provides additional information about the compound. The identification of terpenes in essential oils was aided through the use of different column temperatures [42]. SAAED et al. [43] reported the Kovats indices of twenty monoterpene hydrocarbons on wide bore glass capillary columns coated with SE-30 and Carbowax 20M. Using moderately thick stationary phase films the researchers found only a small influence of column material (soda-lime glass or borosilicate glass) on the retention indices. However, the retention indices were strongly dependent on column wall pretreatment when thin film coatings were used. WRIGHT et al. [44] studied the influence of cross-linking on stationary phase polarity using a variety of free-radical generators. Surprisingly, it was observed that tert.-butyl peroxide caused the greatest increase in column polarity.

Of the polar liquid phases, polyglycol type phases, particularly Carbowax 20M are very popular in flavor research. Their operating properties are inferior to those of the poly-siloxanes; they have relatively high minimum operating temperatures, low maximum operating temperatures and are particularly sensitive to degradation in the presence of

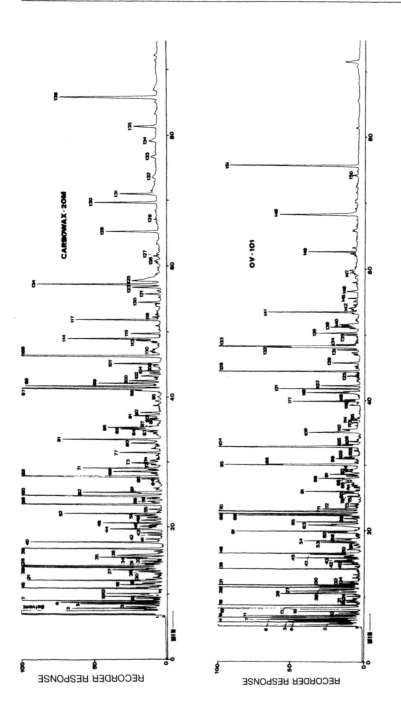

Fig. II.3.–9 Chromatograms of green tea volatiles (obtained by vacuum steam distillation followed by continuous liquid-liquid extraction) analyzed on a 50 m × 0.28 mm i.d. Carbowax 20M column (top) and a 70 m × 0.28 mm i.d. OV-101 column (bottom). The columns were temperature programmed from 80 °C to 220 °C at 2 °C/min [192].

trace levels of oxygen, especially at high temperatures. Oxidation of Carbowax 20M produces trace amounts of acetic acid and acetaldehyde [45]. The later compound can react with amines (1°, 2° and 3°). Therefore, these phases are not recommended for the trace analysis of amines. Commercial "amine deactivated" polyethylene glycol columns which contain small amounts of KOH are available. The surface of fused silica is slightly acidic; the addition of KOH deactivates the column surface through the formation of − Si − OK functionality. The level of KOH must be carefully adjusted so bound KOH exists without the presence of free KOH which would catalyze the breakdown of stationary phase. Due to the hydroscopic nature of KOH these columns have a very limited lifetime.

CONDER and co-workers [46] have studied the thermal degradation of polyethylene glycol 20M and reported that trace levels of oxygen had a critical effect on phase composition. By reducing the concentration of oxygen in the carrier from 10 ppm to 10 ppb the rate of oxidative degradation was decreased by almost five fold and the temperature at which decomposition begins was raised from 160 °C to 200 °C. The importance of proper synthesis and purification of the polyethylene glycols to attain maximum thermal stability has been mentioned [47]. Contamination of the phase with metal ions can catalyze its decomposition. The properties of the Superoxes, polyethylene glycol polymers treated to contain only low levels of residual catalyst have been studied [48, 49]. The gum-like character of the higher Superoxes allowed these workers to prepare columns with very high efficiencies (> 95 %). The Superoxes have a higher operating temperature than Carbowax 20M while maintaining a polarity similar to Carbowax 20M. A drawback of the higher molecular weight Superoxes is their low solubility in most common solvents and the high viscosity of its solutions necessitating the use of a high pressure reciprocating pump for the static coating procedure.

Carbowax 20M is prepared by joining together two poly(ethylene oxide) molecules of MW 6000 by reaction with a proprietary diepoxide [45]. Some disadvantages of Carbowax 20M are its low maximum temperature limit (about 225 °C) and its high minimum temperature limit; at 55−65 °C it undergoes phase transition and solidifies leading to poor chromatographic performance. The lower molecular weight polyethylene glycols such as Carbowaxes 6000, 4000, 1500, 400 have lower minimum operating temperature limits but also posess lower maximum temperature limits than Carbowax 20 M. Another drawback of Carbowax 20M is its wide molecular weight distribution [50]. The polarity of the polyethylene glycol polymer is influenced by its molecular weight. The lower molecular weight polyethylene glycols (such as Carbowaxes 6000, 4000, 1500, 400) have a higher polarity than Carbowax 20 M. The loss of low molecular weight fragments with heating is one source of column bleed and can also lead to changes in column polarity.

The immobilization of polyethylene glycol phase was first reported by DENIJS and DEZEEUW [51]. However, no experimental details were given on the immobilization procedure. TRAITLER et al. [52] described the immobilization of Carbowax 20M with dicumyl peroxide (DCUP) and the coupling agent, γ-glycidoxypropyltrimethoxysilane. The operating range

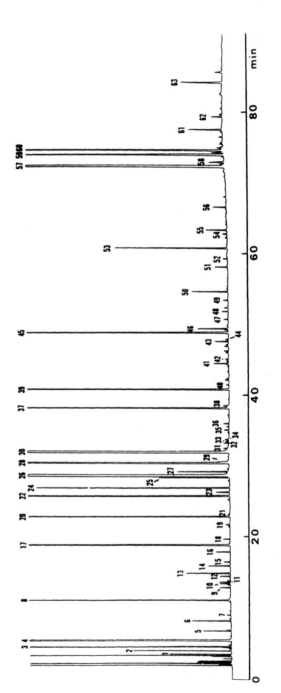

Fig. II.3.–10 Chromatogram of nectarine volatiles (isolated by vacuum steam distillation followed by continuous solvent extraction) analyzed on a 60 m × 0.25 mm i.d. DB-WAX column. The column temperature was programmed from 30 °C (2 min isothermal) to 180 °C at 2 °C/min [171].

of these columns was reported to be 70–280 °C. A major advantage of this procedure is its simplicity. The procedure has been thoroughly evaluated by GROB [53] who proposed several modifications. Sufficient immobilization of the polyethylene glycol was obtained to allow column operation at room temperature. The weakness of this immobilization procedure is the residual column activity. The moderate silylating efficiency of the coupling agent combined with the low reaction temperature result in incomplete silylation of the silica surface [53]. While these columns are very useful for the analysis of fatty acid methyl esters [52] they are probably not the column of choice for the analysis of solutes with free hydroxyl or amine groups. Immobilization of polyethylene glycol 40M with methyl(vinyl) cyclopentasiloxane and DCUP was reported by BUIJTEN et al. [54]. BYSTRICKY [55] used 40 % DCUP to obtain 93–97 % immobilization of Carbowax 20M. The separation of nectarine volatiles on an immobilized polyethylene glycol capillary column is shown is Figure II.3.–10.

The free fatty acid phase, FFAP is produced by esterification of Carbowax 20M with 2-nitroterephthalic acid. The phase has a similar oxygen sensitivity as Carbowax 20M. It has been reported that the phase irreversibly adsorbs aldehydes [56, 57]. The removal of aldehydes appears to be time and temperature dependent and seems to diminish with column use and exposure to air during storage. WITHERS [57] postulated that an acid-catalyzed aldol condensation was responsible for the removal. The dehydration of alcohols by FFAP has been observed by HILTUNEN and RAISANEN [58]. However, these authors reported that FFAP gave the best separation of pine needle oil monoterpene hydrocarbons of nine stationary phases tested. The separation of elder flower extracts using an FFAP capillary column is illustrated in Figure II.3.–11.

The high sensitivity of aldehydes to acid-base effects on all polyglycol phases should be noted since they may tail or fail to elute on neutral or basic columns [59]. Polyglycol columns should be stored under inert gas with sealed ends. The sealing can be done with polyethylene column caps [53], flame sealing or with septa. Besides their oxygen sensitivity, GROB [53] also observed the light sensitivity of the polyglycols. Exposure to daylight (not direct sunlight) resulted in column damage manifested first as an increased bleed rate followed by development of an acidic character. Polyglycol columns should be stored in the dark to avoid these problems.

The polysiloxanes are the most widely used group of stationary phases in gas chromatography. They have been the subject of extensive reviews, noteworthy are those of HAKEN [60, 61] and BLOMBERG [62, 63]. In general, they possess excellent thermal stability and are available in a wide range of polarity. They exhibit good solute diffusivity, even when cross-linked. This accounts for the high efficiency generally observed for columns coated with these phases.

In order to make an inert, efficient and thermally stable capillary column it is necessary to have a well-deactivated surface which can be wetted by the stationary phase. The modi-

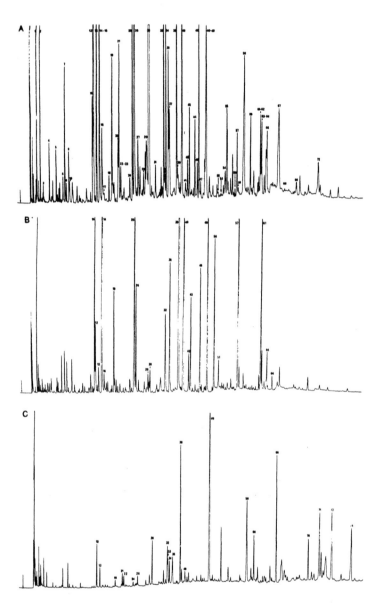

Fig. II.3.–11 Chromatogram of three extracts of elder flowers analyzed on a 60 m × 0.4 mm i.d. FFAP glass capillary column. The column temperature was programmed from 60 °C to 210 °C at 1 °C/min and held. **A** = steam distilled essential oil; **B** = isopentane extract; **C** = ethanol concentrate [172].

fication and deactivation of capillary wall surfaces has been recently reviewed [64] and will not be covered here. A uniform film of stationary phase which does not rearrange when exposed to high temperature is desired in GC. Early attempts to coat polar phases on fused silica were often disappointing. The poor efficiencies and thermal stabilities observed were often attributed to insufficient wettability of the phase onto the column wall. However, using capillary rise measurements BARTLE et al. [65] found that clean fused silica is a high energy surface completely wettable by most stationary phases. Hence raw fused silica should produce efficient columns when coated with most stationary phases. However, another factor influencing stationary phase film stability is the stationary phase viscosity [66]. When exposed to elevated temperatures coated films of low viscosity will be more easily rearranged on the column surface. GOREN [67] has reported that the rate of such rearrangements is inversely proportional to the film viscosity. Methylpolysiloxanes are desirable since they have only a minimal change in viscosity with temperature [68]. This has been attributed to their chemical structure; the methylpolysiloxane molecule reportedly has a helical structure in which the siloxane bonds, oriented toward the axis are shielded by the outwardly pointing methyl groups [69]. A rise in temperature increases the mean intermolecular distance, causing a decrease in viscosity; this decrease in viscosity is partly offset by expansion of the helices, thus decreasing the mean intermolecular distance. The result of these two opposing actions is that the net mean intermolecular distance is only slightly changed with temperature and thus there is little change in the viscosity. The substitution of larger groups, such as cyanopropyl or phenyl disrupts the helical structure, resulting in a greater change in viscosity with temperature. Stationary phases whose viscosities become low at elevated temperatures are more susceptible to rearrangements on the column surface than phases whose viscosities change only slightly with temperature. This leads to reduced column efficiency due to uneven film distribution. The more viscous gum methylpolysiloxanes undergo a smaller relative change in viscosity with temperature than the less viscous fluid methyl polysiloxanes and hence gum phases are preferred over fluids.

PEADEN et al. [70] recently prepared 50 and 70 % phenylmethylphenylpolysiloxanes gum phases that were more viscous than commercially available phenyl phases and produced thermally stable and efficient fused silica columns. The incorporation of 1–4 % vinyl groups in the phases facilitated cross-linking resulting in improved film stability. Immobilization of the phase through in situ cross-linking represents a major breakthrough in capillary chromatography. Stabilization of the stationary phase film is attained thereby minimizing column deterioration due to film rearrangement on the capillary surface at elevated temperatures. This stabilization is particularly important for the preparation of efficient polar capillary columns. Stationary phase immobilization not only improves the thermal stability but extends the lower temperature limits as well. The separation of C1-C5 hydrocarbons on fused silica columns coated with cross-linked and non-cross-linked methylpolysiloxane gum has been compared [71]. The non-cross-linked columns suffered a sizable loss of efficiency at $-60\,°C$ and below while the cross-linked columns showed a more gradual loss of efficiency. Cross-linking of the stationary phase apparently suppresses the glass transition mechanism. Another advantage of immobilization is that the stationary phase is rendered insoluble by most solvents. Large volume on-column or splitless injections can thus be done without the problem of phase stripping. BLOMBERG and co-workers [72] have demonstrated the ability of cross-linked capillary columns to withstand aqueous injections. The introduction of water by direct injection or headspace sampling should be done with care since water causes retention time shifts [73, 74] and can extinguish the flame of an FID. The injection of acids and bases together with water can lead to acid or base catalyzed hydrolysis of the stationary phase. This generates terminal silanol groups, which catalyze the more rapid decomposition of the polymer chain. The hydroxyl group attacks the polymer chain to release polysiloxane rings of various sizes. This release is manifested as increased column bleed rates. The degradative reactions can be blocked by capping the exposed hydroxyl groups. An effective method of resily-

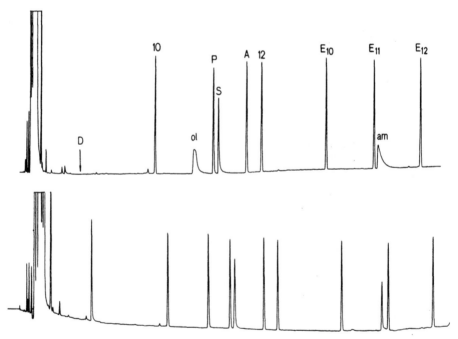

Fig. II.3.–12 Grob test chromatograms on a used 20 m × 0.33 mm i.d. SE-54 (d_f – 2.5 μm) column before (top) and after (bottom) re-silylation. The improved elution of alcohols, **D** = 2,3-butanediol; **ol** = 1-octanol, is noteworthy [75].

lation to reduce bleed rates has been detailed by GROB and GROB [75]. The authors first recommend solvent rinsing the column which necessitates the use of immobilized stationary phases. The solvent washing step is followed by resilylation with diphenyltetramethyldisilazane (DPTMDS) which reduces the bleed rate of both new and used columns. Used columns may display improved inertness due to a cleaning effect where high molecular weight polar contaminants are transformed into more soluble species (Fig. II.3.–12). JENNINGS [76] has confirmed the effectiveness of this silylation procedure. Finally, cross-linking makes possible the production and use of thick film columns ($d_f > 1-2$ µm).

Ideally, the immobilization step should not change the stationary phase characteristics. WRIGHT et al. [44] has studied the effect of free radical cross-linking (using various peroxides and azo compounds) on the polarity and inertness of SE-54 capillary columns. The major problem occurring with free radical cross-linking has been the increased column activity resulting from free radical generator decomposition products. The cross-linked columns exhibited increased octanol retention and greater adsorptive activity toward alkyl amines. BEREZKIN and KOROLEV [77] also noted an increased retention of polar compounds on cross-linked phases compared to non-cross-linked phases. They attributed this behavior to an increased contribution to adsorption to retention. The methods of stationary phase immobilization have been reviewed by HAKEN [60] and BLOMBERG [78] and will not be discussed here.

Methylpolysiloxanes are the lowest polarity polysiloxanes. These phases form the most uniform and stable films on glass and fused silica resulting in the highest column efficiencies. Their excellent inertness, wide temperature operating range (– 60 °C to 350 °C) and relative durability explain their wide use in gas chromatography. These phases should be tested first when selecting a capillary column for a specific application. The interaction between solutes and methylpolysiloxanes are largely dispersion forces; solute elution is based on their vapor pressure and, in general, solutes elute according to their boiling point.

The trifluoropropyl group has a high dipole moment and a strong electron acceptor ability which results in the unique selectivity of this phase. Fluoropolysiloxanes are less heat stable than the methylpolysiloxanes due to their susceptibilty to oxidative degradation and siloxane bond cleavage [79]. The effectiveness of the trifluoropropyl phase in the separation of chiral flavor components as their diastereomeric derivatives has been demonstrated. TRESSL and ENGEL [80] prepared diastereomeric R-(+)-α-methoxy-α-trifluoromethylphenylacetic acid (R-(+)-MTPA) derivatives of various 3-hydroxy acid esters and found the best separation of R-(+)-MTPA derivatives of ethyl 3-hydroxybutanoate was obtained on a DB-210 (50 % trifluoropropyl, 50 % methylpolysiloxane) column. TRESSL and co-workers [81] reacted chiral secondary alcohols with R-(+)-phenylethylisocyanate (R-(+) PEIC) to form diastereomeric urethane derivatives. High relative retention, α values for these derivatives were obtained on a DB-210 column.

GROB, Jr. and GROB [59] discussed the practical significance of varying the film thickness and gave suggestions for selecting the proper thickness for different types of samples. Thick film columns of this time had very short lifetimes. The production of stable thick filmed columns became possible when cross-linking and bonding technology developed. Today, commercial capillary columns are available in film thickness (d_f) ranging from 0.1 to 8.0 μm. The typical average film thickness is 0.25 μm. This represents a good compromise in terms of efficiency, solute capacity and retention. Increasing the film thickness results in larger solute partition ratios (k). At the same column diameter doubling the film thickness will halve phase ratio, β ($\beta = V_g/V_l$, V_g = column gas phase volume, V_l = column stationary phase volume), resulting in a doubling of k. Thick film columns ($d_f > 0.5$ μm) are useful for the analysis of highly volatile compounds such as in a headspace sample. The solute capacities also dramatically increase with increasing film thickness. This may confer advantages when using techniques such as gas chromatography – Fourier transform infrared spectroscopy (GC-FTIR) which require larger sample amounts. Thick film columns should be used with caution since they are not desirable for many applications. Thick film columns exhibit lower efficiencies than columns coated with conventional film thicknesses ($d_f = 0.25$ μm). As d_f increases above about 0.4 μm column efficiencies begin to decrease due to an increase in the Cs term in the van Deemter equation. There is a higher level of bleed associated with thick film columns. This can pose problems in GC/MS applications.

For the analysis of low volatility or thermally labile compounds thin film columns ($d_f = 0.15$ μm) should be considered. Compounds will be eluted at shorter retention times than on normal or thick film columns. In temperature programmed operations compounds will be eluted at lower temperatures all other factors being same. CRAMERS et al. [82] concluded that in most situations thin film columns are preferred over thick film columns in trace analysis.

4   Prefractionation

4.1   Column Chromatography on Silica Gel

The identification of minor constituents in complex flavor mixtures is greatly facilitated by the preseparation into different fractions prior to gas chromatographic analysis. In the analysis of essential oils, the separation of the oxygenated fraction from the terpene hydrocarbons can greatly aid the identification of trace constituents. Column chromatography using silica gel is a widely used and effective method of fractionation. The hydrocarbons can be readily separated from the total oil by selective elution with pentane or hexane [83]. The polar, oxygenated constituents can then be eluted using gradient elution with increasing levels of diethyl ether. The application of this method has aided in the identification of the photocitral isomers in natural verbena oil [84]. A silica gel-aluminum oxide (2:1) column was used by TRESSL et al. [85] to separate a beer extract into six fractions. The pretreatment led to the identification of 45 new beer constituents. The identification of a number of hop oil constituents was noteworthy since these components are normally masked by other beer constituents. SCHREIER and co-workers have utilized silica gel prefractionation with a pentane-diethyl ether gradient in their examination of the volatiles from cherimoya [86], papaya [87], guava [88] and Alphonso mango [89].

Silica gel must be used with caution since it may cause compositional changes in samples due to isomerization reactions. The isomerization of limonene with silica gel at 100 °C and 150 °C has been studied [90]. Limonene initially isomerized to α-terpinene, γ-terpinene, terpinolene and isoterpinolene, which subsequently underwent disproportionation and polymerization reactions. The isomerization of sabinene into α-thujene, α-terpinene, γ-terpinene, limonene, β-phellandrene and terpinolene during thin-layer chromatography has been reported [91]. Evidence supported an acid-catalyzed isomerization of sabinene since silica gel chromatostrips prepared by using 0.1 N NaOH in place of distilled water resulted in much lower isomerization. It has been suggested that acid-catalyzed reactions observed with silica gel chromatography are probably due to presence of acidic impurities since carefully purified silicas exhibit only small tendency toward acid-catalyzed sample reactions [92].

To prevent the isomerization of sensitive terpenes during column chromatography, stationary phases such as Emulphor-O [93] and Carbowax 20M [94] have been used to deactivate active sites on silica gel. However, the rearrangement of citronellal has been reported despite the deactivation of silica gel with polyethylene glycol [95]. Additionally, since the stationary phases are not bonded to the silica they may be eluted from the silica and contaminate the various fractions.

The use of water is an effective and simple alternative for deactivation of silica gel. No changes in a monoterpene hydrocarbon mixture were found after column chromatography on silica gel with a water content of 7 % [96]. SCHEFFER et al. [97] were able to prevent the isomerization of various monoterpene hydrocarbons by using an acid and base washed silica gel with a 5 % water content. They reported that the water content of the silica gel should be kept low for the efficient separation of the terpenes.

Despite the use of other chromatographic adsorbents such as Florisil [98] and alumina [99], silica is preferred due to its greater linear capacity and higher column efficiencies.

4.2 HPLC

High performance liquid chromatography (HPLC) is being increasingly used for the prefractionation of flavor mixtures. Due to its greater speed and efficiency, HPLC has advantages over conventional column chromatography. There is less danger of sample rearrangements since the material is in contact with the adsorbent for a much shorter time. The sample throughput is increased and the improved resolution is advantageous in the subsequent GC analyses. The use of HPLC in the isolation and analysis of flavor constituents has been reviewed by BITTEUR [100]. MORIN et al. [101] discussed the separation of terpene compounds using the main modes of HPLC (adsorption, reversed-phase, exclusion, ligand exchange). A strategy using various separation techniques to analyze complex flavor mixtures was proposed by TEITELBAUM [102]. The first step involved a preliminary clean-up with a silica gel column. The next two steps involved the use of HPLC, first in the adsorption mode and then in the partition mode (normal or reversed phase). In the final step, GC is used to separate the desired fractions. Many of the early flavor fractionating studies utilizing HPLC involved the use of synthetic mixtures. JONES et al. [103] used HPLC to separate a model mixture of terpenes. In their study of a model browning system, YAMAGUCHI et al. [104] used HPLC to fractionate a complex mixture of heterocyclic compounds (furans, thiophenes, pyrroles, thiazoles, oxazoles, pyrazines and imidazoles) prior to GC analyses. Reversed-phase HPLC was utilized by KUBECZKA [105] to fractionate a mixture of oxygenated terpenoids, monoterpene hydrocarbons and sesquiterpene hydrocarbons in 12 min. A general HPLC method for the prefractionation of oxygenated constituents of essential oils was proposed by CHAMBLEE et al. [106]. Samples were fractionated on three different silica columns in series using a ternary solvent system of ethyl acetate:methylene chloride:hexane (8:46:46, v/v). Collected fractions were concentrated and then directly analyzed by GC-MS. The application of this technique to cold pressed lime oil aided the identification of 23 new constituents. MORIN et al. [107] used reversed-phase HPLC to separate essential oil constituents before applying GC-MS. The Vervain essential oil was separated on a C_{18} phase using a water-acetonitrile gradient. The use of aqueous solvents in reversed-phase systems is a limitation since fractions must be extracted before they can be concentrated and analyzed by GC.

The isolation of germacrene B, an important flavor impact compound of lime peel oil, was accomplished through the use of a combination of chromatographic techniques [108]. The separation scheme is shown in Figure II.3.–13. Column chromatography with silica gel was used to separate the hydrocarbon compounds in the residue. The hydrocarbon fraction was subjected to preparative GC using a 12 ft. X ¼ in. (i.d.) glass column packed with HP-5 % Triton X-305 on Chromosorb W (80–100 mesh). The partially purified germacrene B was finally cleaned up using normal phase HPLC.

4.3 Preparative GC

Preparative GC is another traditional method for the prefractionation and purification of flavor mixtures. The technique must be used with care since there is the risk of thermal degradation of sensitive compounds. Packed columns have been typically used in preparative applications though systems designed for micropreparative isolation and enrichment from capillary columns have been described [109, 110]. One limitation of packed columns is the activity of the support material. As the stationary phase loadings on the support decrease, the inertness of the support becomes increasingly important. Acidic

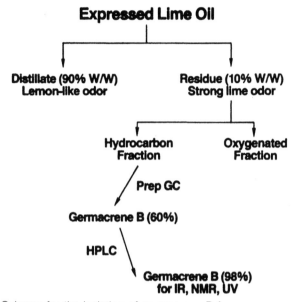

Fig. II.3.–13 Scheme for the isolation of germacrene B from expressed lime oil [108].

sites on the support can cause the isomerization and decomposition of sensitive constituents. GILLEN and SCANLON [111] reported that the enolization of isomenthone increased with decreasing stationary phase loadings. DEMOLE et al. [112] stressed the importance of column deactivation in their isolation and identification of the important grapefruit flavor impact compound, 1-p-menthene-8-thiol. The researchers used semipreparative GC in one of their isolation steps. It was necessary to deactivate the packed silicone column by treatment with dimethylaminoethanol and Silyl-8 (Pierce Chemical Company, Rockford, IL.) to elute the sulfur compound. Glass columns are preferred over metal columns for reasons of inertness. DEBRAWERE and VERZELE [113] used preparative GC (glass column 12 m × 11 mm i.d., 10 % Carbowax 20 M on Chromosorb G, 30–40 mesh) to isolate various monoterpene hydrocarbons from black pepper oil. They subsequently obtained structural confirmation of the collected terpenes by IR and NMR spectroscopy. Sample decomposition in the thermal conductivity detector (TCD) may occur in preparative GC [114, 115]. VERZELE [115] reported that an acid contaminated TCD can cause sample isomerization. As an alternative, a glass effluent splitter may be used at the column exit with 2–5 % of the flow directed to an FID [116].

4.4   Preconcentration of Flavor Precursors

Flavor precursors typically exist as complex mixtures of polar constituents which are often conjugated as glycosides. Their water soluble nature makes them difficult to separate from other highly polar material such as sugars and organic acids. An elegant solution to this difficult analytical problem was provided by WILLIAMS and co-workers [117] who used a C_{18} reversed-phase adsorbent to isolate monoterpene glycosides and nor-isoprenoid precursors from grape juice and wines. These compounds are selectively retained on the adsorbent (other polar material such as sugars and organic acids are washed off the column with water) and a 20,000 fold concentration can be achieved in a single chromatographic step. The method has been subsequently used to examine the glycosidically bound components of yellow passion fruit [118], apricot and mango [119]. Additionally, the non-ionic polystyrene copolymer, Amberlite XAD-2, has been used to isolate glycosidic components in grape juice and wines [120].

Droplet countercurrent chromatography (DCCC) has been applied to the resolution of polar constituents in grape juice [121, 122]. The technique aided in the separation of the conjugated forms, including glycosides of various terpenoids and phenols. Juice samples were prepared by prefractionation on C18 reversed-phase adsorbent or were directly subjected to DCCC.

4.5   Headspace Sampling

This topic has been the subject of many reviews [123, 124] and books [125, 126]. The reader is referred to these references for background and specialized information. The present review will examine new trends in headspace sampling and discuss their role in flavor research.

4.5.1   Direct Headspace Injection

Direct analysis of headspace vapors offers several advantages; the technique is simple, rapid and minimizes qualitative and quantitative changes that often occur in other sample preparation methods. The method involves little or no sample preparation. In static headspace sampling the sample is placed in a closed container and allowed to come to thermodynamic equilibrium. A sample of headspace vapor is then transferred to the GC system. In dynamic systems there is a continual gas flow swept through the container holding the sample (Fig. II.3.–14). This method may be preferred for following changes in living tissues (e.g. following changes during fruit ripening). Because of the limited concentration of volatiles in the headspace both methods usually require relatively large sample volumes which leads to the introduction of long sample bands and results in broad chromatographic peaks. The use of on-column injection permits the direct introduction of volatiles inside a fused silica column thus avoiding the dilution by carrier gas normally involved in conducting the sample from the injection port to the column. Cryogenic focusing is normally employed to shorten sample bands resulting in enhanced chromatographic performance. Cryogenic focusing involves the use of subambient temperatures over a portion or all of the column to concentrate the sample in a narrow band at the head of

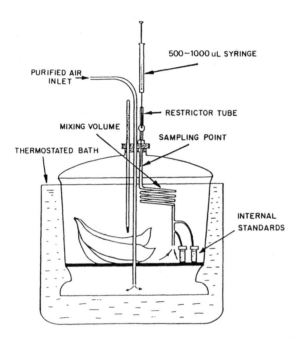

500–1000 uL SYRINGE

PURIFIED AIR INLET

RESTRICTOR TUBE

MIXING VOLUME

SAMPLING POINT

THERMOSTATED BATH

INTERNAL STANDARDS

Fig. II.3.–14 Headspace sampling chamber for following changes in ripening fruit volatiles [173].

the column. TAKEOKA and JENNINGS [127] have discussed the theory of cryogenic focusing and have described a simple procedure for focusing on-column headspace injection (Fig. II.3.–15). The authors place a short loop (25 cm) of the capillary column into a Dewar flask containing liquid nitrogen at the time of injection and remove it to commence the chromatographic run. The procedure has been applied to the analysis of ripening bananas [128], birch syrup [129] and onions [130] and smoked sausages and bacon [131]. A chromatogram of smoked sausage (Landjaeger) headspace volatiles is shown in Figure II.3.–16.

The size of the headspace sample is usually restricted by the formation of the ice plug that takes place inside the column. However, the use of wide-bore columns (0.5–0.7 mm i.d.) has permitted the analysis of air sample volumes as large as 100 mL [132]. Additionally, wide bore pre-columns can be attached to the front of conventional capillary columns (0.25–0.32 mm i.d.) to accomodate large volume headspace samples without sacrificing column efficiency.

Water may cause other problems in direct headspace sampling. BURNS and co-workers [73] have shown that water vapor shifts solute retention times (1–5 %) on fused silica

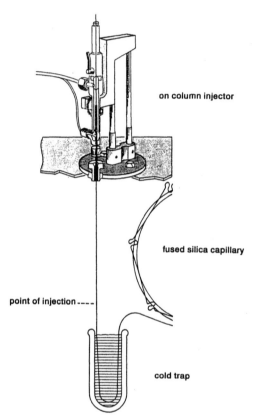

Fig. II.3.–15 Schematic of a system for on-column headspace injections.

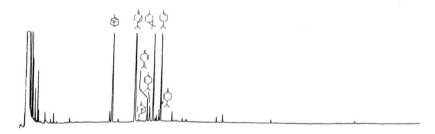

Fig. II.3.–16 Chromatogram produced by on-column headspace injection (10 ml) of smoked meat sample (Landjaeger). A 30 m × 0.25 mm i.d. DB-1 column was employed [131].

columns. Retention times of solutes increase with increasing water content until maximum retention shifts are obtained. The mechanism of this effect is not understood since the non-polar DB-1 column used has little affinity for water. It has been observed that water forms clusters on methylpolysiloxane films [133] and the resulting aggregates may influence solute retention. In the analysis of birch syrup headspace volumes up to 15 mL could be injected without a shift in retention times [134]. GROB and HABICH [74] studied the effect of much larger amounts of water (split injections of solutions containing 30–100 % water) on solute retentions on a polyethylene glycol column. Retained water caused an increase in column polarity; non-polar solutes were eluted much earlier while polar solutes displayed longer elutions. The magnitude of the effect was influenced by the amount of water introduced into the column. To obtain reproducible results with headspace and direct aqueous injections the column should be heated above 100 $^\circ$C between runs to remove adsorbed water. The large amount of water may occasionally extinguish the flame of an FID during a chromatographic run. This problem can be alleviated by increasing the air flow rate to 600 mL/min [135].

Another complication with direct headspace injections is the introduction of large amounts of oxygen into the column. The premature deterioration of polyethylene glycol columns with 10 mL headspace injections has been observed [134]. While it may be useful to pre-purge the sample container with an inert gas such as nitrogen or helium the use of more oxygen resistant phases such as methylpolysiloxane or methyl-phenylpolysiloxane is recommended.

With regard to the injection speed with gas-tight syringes, it has been found that slow injections (0.5 mL/min) favor the recovery of the more volatile solutes [134]. This is probably due to the higher inlet pressures generated with fast injection speeds which may force solute vapors back up the carrier line or out the top of the on-column injector in the space between the fused silica needle and the restrictor tube. Similar effects have been noted for splitless headspace injection. Fast splitless injections produced lower solute recoveries since a large portion of the sample was forced into the septum purge system and exited out the vent line [136].

4.5.2 Trapping with Very Thick Film Columns

The development of cross-linking and bonding technology has permitted the production of very thick (12–100 μm) film columns. These columns represent an alternative to classical adsorbents such as Tenax or charcoal in headspace trapping. Cross-linked stationary phases offer temperature stability, inertness and compatibility with most solvents (useful when using solvent elution). In addition, no displacement effects as encountered with solid adsorbents occur since retention is based on dissolution. GROB and HABICH [137] pioneered the use of thick film capillary traps in headspace analysis. The authors tested short (60 mm × 0.3 mm i.d.) traps coated with 12–15 μm immobilized PS-255. The traps possessed excellent inertness when tested with heat labile compounds. The biggest drawback of the traps was their low retention. BURGER and MUNRO [138] studied the breakthrough volumes of 1 m (0.32 mm i.d.) capillary traps with film thicknesses of 12 µm. Breakthrough volumes of 0.2, 0.4, 0.9, 3.5, 13 and 26 mL were obtained for propane, methanol, pentane, chloroform, toluene and octane, respectively. These values again confirm the low retention on fused silica traps. BICCHI et al. [139] used longer (up to 3 m) capillary traps coated with 15 µm stationary phase films to collect volatiles emitted by living plants. The authors found quantitative and qualitative differences between sampling under pressure and sampling under vacuum. In general, breakthrough volumes were lower when sampling under vacuum. In contrast to previous studies the authors used intermediate cryogenic focusing of desorbed volatiles prior to sequential analysis on an analytical column.

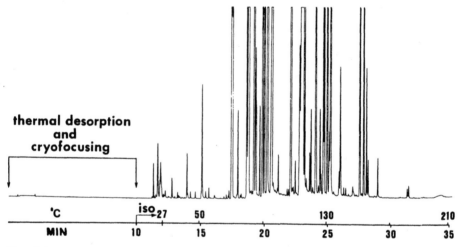

Fig. II.3.–17 Chromatogram of spice (Origanum vulgare) headspace volatiles collected on a thick film capillary trap (1.5 m × 0.7 mm i.d., d_f – 91 µm) coated with cross-linked Sylgard 184. The volatiles were desorbed from the trap at 100 °C and cryofocused on a cross-linked OV-1701 capillary column (20 m × 0.32 mm i.d., d_f – 0.23 µm) [140].

To overcome the problem of low retention on capillary traps a new coating method was developed to allow the production of stationary phase films up to 100 µm thick [140]. The method involves an immediate cross-linking of a prepolymer film formed during static coating. The resulting columns have low phase ratios and hence exhibit very high retention. The retention of such a thick film column (d_f = 80 µm) is roughly equivalent to the retention of an empty tube held at a temperature 80–90 °C lower. With these thick film traps it is possible to trap headspace samples or effluent from capillary columns for preparative work. The volatiles can then be released from the trap by either thermal desorption which yields a solvent free extract or solvent elution which provides milder conditions for heat labile compounds. A chromatogram of spice headspace volatiles obtained by concentrating volatiles on a thick film trap followed by thermal desorption to the analytical column is shown in Figure II.3.–17.

5   Important Flavor Compounds

The compounds included in this category possess unique sensory properties (Fig. II.3.–18). Though foods and essential oils contain complex mixtures of chemicals the odor character of a food or oil may reside largely in one or a few compounds. Such compounds have been termed "character impact compounds" by JENNINGS and SEVENANTS [141]. Examples include 2-isobutyl-3-methoxypyrazine which has the characteristic odor of green bell pepper and ethyl (E, Z)-2,4-decadienoate which possesses the typical Bartlett pear odor. These impact compounds are typically potent odorants possessing very low odor thresholds. The odor thresholds reported in Figure II.3.–18 were measured in water. The grapefruit impact compound, 1-p-menthene-8-thiol, and bis (2-methyl-3-furyl) disulfide have the lowest known thresholds. The latter compound had been previously identified as a constituent produced from heating a mixture of thiamin hydrochloride, L-cysteine, hydrolyzed vegetable protein and water [142]. The authors also identified the related compound, 2-methyl-3-furanthiol in the mixture. This compounds which has a strong beef extract aroma was recently found in canned tuna fish [143]. 3,9-Epoxy-p-menth-1-ene is the impact compound of fresh dill herb. BRUNKE and ROJAHN [144] have concluded that its absolute configuration must be that shown in Figure II.3.–18. The hydrocarbon, 1-(E, Z)-3,5-undecatriene, has been identified as an impact compound of the essential oil of Galbanum [145, 146]. 1-(E, Z,Z)-3,5,8-Undecatetraene first identified in the essential oil of the Hawaiian seaweed Dictyopteries [147] acts as a sperm attractant in the brown alga Ascophyllum nodosum [148]. This hydrocarbon was later found in mango by IDSTEIN and SCHREIER [149]. These unsaturated hydrocarbons probably play an

1-p-menthene-8-thiol	3,9-epoxy-p-menth-1-ene	2-isobutyl-3-methoxypyrazine
grapefruit	dill herb	bell pepper
threshold - (+)-(R)- 2×10^{-5} ppb	[174]	threshold - 0.002 ppb
(-)-(S)- 8×10^{-5} ppb		[175]
[112]		

1-(E,Z)-3,5-undecatriene

1-(E,Z,Z)-3,5,8-undecatetraene

pineapple

[150, 151]

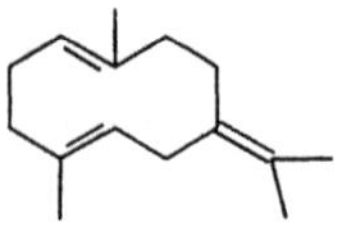

germacrene B

Key lime peel oil

[108]

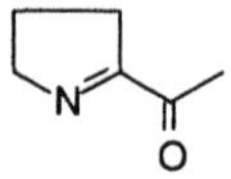

2-acetyl-1-pyrroline

cooked rice

threshold - 0.1 ppb

[152]

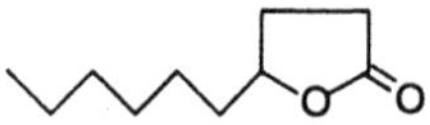

γ - decalactone

peach

[141]

threshold - 0.7 ppb

[181]

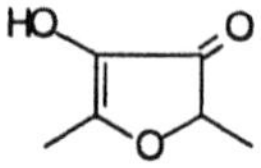

2,5-dimethyl-4-hydroxy-3(2H)-furanone

(furaneol)

pineapple

[176]

threshold - 0.03 ppb

[177]

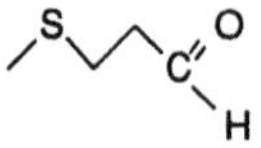

3-methylthiopropanal

cooked potato

[178]

threshold - 0.2 ppb

[179]

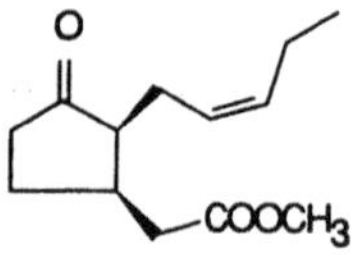

1R, 2S (+)-Z-methyl epijasmonate

commercial methyl jasmonate

threshold - 3 ppb

[157]

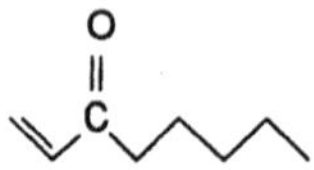

1-octen-3-one

boiled mushroom

[180]

threshold - 0.005 ppb

[181]

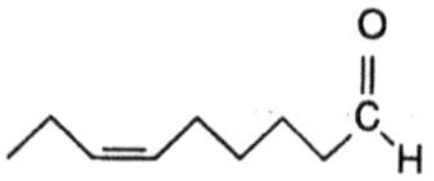

(Z)-6-nonenal

melon

[159]

threshold - 0.005 ppb

[182]

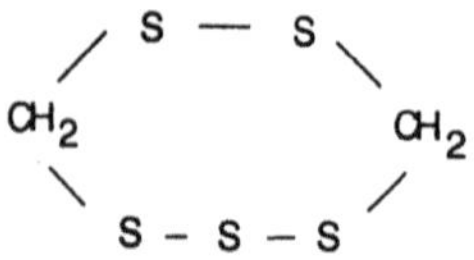

lenthionine

1,2,3,5,6-pentathiepane

Shiitake mushroom

[183]

threshold - 400 ppb

[184]

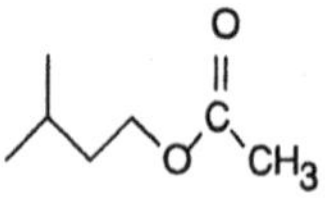

3-methylbutyl acetate

banana

[185]

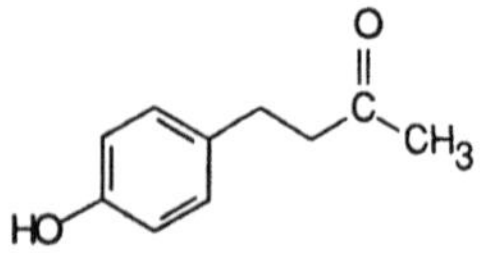

"raspberry ketone"

4-(4-hydroxyphenyl)-2-butanone

[186]

threshold - 100 ppb

[187]

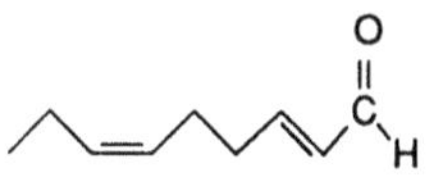

(E,Z)-2,6-nonadienal

cucumber

[160]

threshold - 0.01 ppb

[181]

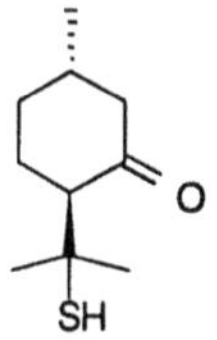

(+)-trans-8-mercapto-p-menthan-3-one

buchu leaf oil

[161]

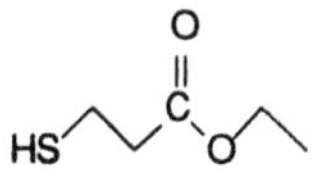

ethyl 3-mercaptopropanoate

Concord grape

threshold - 0.2 ppm

[163]

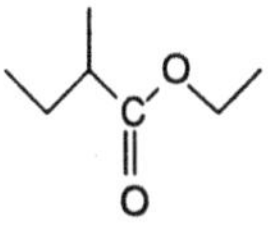

ethyl 2-methylbutanoate

Delicious apple

threshold - 0.3 ppb

[188]

$CH_3 - S - S - S - CH_3$

dimethyl trisulfide

cabbage, broccoli, cauliflower

threshold - 0.01 ppb

[29]

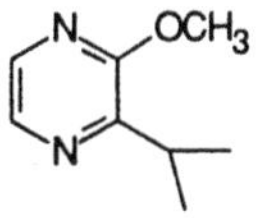

2-methoxy-3-isopropylpyrazine

potato

[164]

threshold - 0.002 ppb

[189]

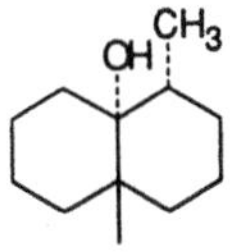

geosmin

(trans-1,10-dimethyl-trans-9-decalol)

beetroot

[165]

threshold - 0.02 ppb

[181]

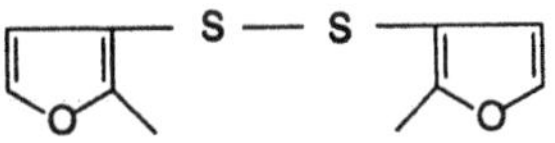

bis(2-methyl-3-furyl) disulfide

"thiamin odor compound"

threshold - 2×10^{-5} ppb

[190]

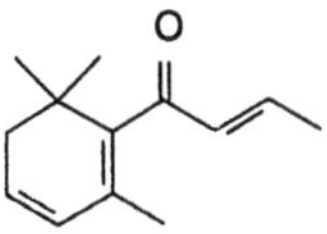

β - damascenone

Bulgarian rose oil

[167]

threshold - 0.002 ppb

[170]

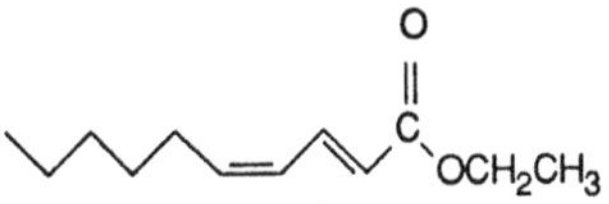

ethyl (E,Z)-2,4-decadienoate

Bartlett pear

[191]

Fig. II.3.–18 Important Flavor Compounds.

important role in pineapple flavor [150]. Though the thresholds of these hydrocarbons have not been reported, sniffing of the gas chromatographic effluent permitted detection in the picogram range [150]. These potent odorants have been found in various fruits and vegetables [151]. The double bond configuration in the 5 position is crucial to the odor potency; the corresponding isomers, 1-(E, E)-3,5-undecatriene and 1-(E, E, Z)-3,5,8-undecatetraene have odor thresholds 10^6 and 10^4 times higher, respectively [150]. These hydrocarbons must be analyzed with care since they are sensitive to enzymatic and/or oxidative degradation [150]. The sequiterpene hydrocarbon, germacrene B, has a warm, sweet, woody-spicy, geranium-like odor and is an important compound in expressed lime oil [108]. The authors reported a 10–15 % loss (relative to HPLC) of this labile sesquiterpene when analyzed by GC. The important cooked rice aroma compound, 2-acetyl-1-pyrroline, particularly characterizes the more aromatic varieties such as Basmati where it occurs at high concentrations [152]. This compound was later found to be the dominant odor constituent of wheat bread crust [153]. Due to its low odor threshold this "crackerlike" odor compound probably has an important flavor role in other baked products. γ-Decalactone is an important aroma constituent in peach and nectarine [154]. 2,5-Dimethyl-4-hydroxy-3(2H)-furanone (furaneol) has been identified in a variety of foods [155]. At higher concentrations furaneol has a caramel, burnt sugar odor while at lower concentrations it beomes fruity and strawberry-like [156]. Odor studies of the four Z stereoisomers of methyl jasmonate revealed that only (+)-methyl epijasmonate has a strong and characteristic odor and thus is largely responsible for the odor of commercial mixtures of methyl jasmonate [157]. The important role of 1-octen-3-one in fresh mushroom odor was recently confirmed by FISCHER and GROSCH [158]. KEMP et al. [159] identified a key melon constituent, (Z)-6-nonenal in muskmelon. Lenthionine is a significant compound in the flavor of Shiitake mushroom. The identification of the cucumber impact compound, (E, Z)-2,6-nonadienal was a noteworthy achievement since it pre-dated the use of gas chromatography [160]. Two diastereomeric forms of p-methane-8-thiol-3-one-occur at the level of about 0.5 % in Buchu leaf oil [161]. The trans isomer imparts the black current principle in Buchu leaf oil [162]. Ethyl 3-mercaptopropanoate, a Concord grape component, has a skunky or foxy aroma at high concentrations while on dilution it takes on a pleasant fruity, grape-like character [163]. Its contribution to the "foxy" character of some American wines has been postulated [163]. Based on the amounts present and its low odor threshold dimethyl trisulfide is considered to be an important component of cooked cabbage, broccoli and cauliflower [29]. 2-Methoxy-3-isopropylpyrazine has been identified in potato where it contributes to the earthy aroma [164]. Though implicated in musty-earthy off-odor taints in water supplies geosmin is an important constituent of beetroot [165] and also contributes to the earthy aroma of soil [166]. The identification of the important Bulgarian rose oil constituent, β-damascenone was reported by DEMOLE et al. [167]. This potent odorant has also been detected in apple [168], grape [169] and tomato [170].

References

1. MITZNER, B. (1964) Anal. Chem. **36**: 242

2. GROB, K., GROB, G. (1979) J. High Res. Chromatogr. **2**: 109

3. CARDOSO, J. N., ALFONSO, J. C. (1988) J. High Res. Chromatogr. **11**: 537

4. GROB, K., GROB, K., Jr. (1978) J. Chromatogr. **151**: 311

5. ROERAADE, J., FLODBERG, G. BLOMBERG S. (1985) J. Chromatogr. **322**: 55

6. GROB, K., Jr., GROB, G., GROB, K. (1978) J. Chromatogr. **156**: 1

7. GROB, K. (1978a) J. High Res. Chromatogr. **1**: 263

8. SCHOMBURG, G., BEHLAU, H., DIELMANN, R., WEEKE, F., HUSMANN, H. (1977) J. Chromatogr. **142**: 87

9. GROB, K., Jr., NEUKOM, H. P. (1980) J. Chromatogr. **189**: 109

10. GROB, K. (1987) in "On-Column Injection in Capillary Gas Chromatography", Huethig Verlag, Heidelberg, Basel, New York

11. JENNINGS, W. (1987) in "Analytical Gas Chromatography", Academic Press, Inc. New York

12. GROB, K., Jr. (1978b) J. High Res. Chromatogr. **1**: 307

13. BERTHOU, F., DREANO, Y. (1979) J. High Res. Chromatogr. **2**: 251

14. GROB, K., Jr. (1982) J. Chromatogr. **237**: 15

15. GROB, K., MÜLLER, E. (1987) J. Chromatogr. **404**: 297

16. GROB, K., Jr., FRÖHLICH, D., SCHILLING, B., NEUKOM, H. P., NAGELI, P. (1984) J. Chromatogr. **295**: 55

17. ROHWER, E. R., PRETORIUS, V., APPS, P. J. (1986) J. High Res. Chromatogr. **9**: 295

18. BRETSCHNEIDER, W., WERKHOFF, P. (1988) J. High Res. Chromatogr. **11**: 543

19. VOGT, W., JACOB, K., OBWEXER, H. W. (1979) J. Chromatogr. **174**: 437

20. POY, F., VISANI, S., TERROSI, F. (1981) J. Chromatogr. **271**: 81

21. SCHOMBURG, G., HUSMANN, H., BEHLAU, H., SCHULZ, F. (1983) J. Chromatogr. **279**: 251

22. GROB, K. Jr., NEUKOM, H. P. (1979) J. High Res. Chromatogr. **2**: 15

23. GROB, K., LAULDI, T., BRECHBUEHLER, B. (1988) J. High Res. Chromatogr. **11**: 462

24. NITZ, S., DRAWERT, F., JULICH, E. (1984) Chromatographia. **18**: 313

25. WERKHOFF, P., BRETSCHNEIDER, W. (1987) J. Chromatogr. **405**: 99

26. BERTSCH, W., SHUNBO, F., CHANG, R. C., ZLATKIS, A. (1974) Chromatographia. **7**: 128

27. PRETORIUS, V., ROHWER, E. R., HULSE, G. A., LAWSON, K. H., APPS, P. J. (1984) J. High Res. Chromatogr. **7**: 429

28. FLATH, R. A., FORREY, R. R. (1970) J. Agric. Food Chem. **18**: 306

29. BUTTERY, R. G., GUADAGNI, D. G., LING, L. C., SEIFERT, R. M., LIPTON, W. (1976) J. Agric. Food Chem. **24**: 829

30. AVERILL, W., MARCH, E. W. (1976) Chromatogr. Newsl. **4**: 20

31. BLOMBERG, L., BUIJTEN, J., MARKIDES, K., WANNMANN, T. (1983) Proc. 5th Int. Symp. on Capillary Chrom. (Rijks, J., Ed.) Elsevier, Amsterdam, p. 99

32. JENNINGS, W. (1981) "Comparisons of Fused Silica and Other Glass Columns in Gas Chromatography", Huethig Verlag, Heidelberg

33. LEE, M. L., YANG, F. J., BARTLE, K. D. (1984) "Open Tubular Column Gas Chromatography: Theory and Practice", Wiley-Interscience. New York

34. DESTY, D. H. (1987) J. Chromatogr. Sci. **25**: 552

35. TRESTIANU, S., ZILIOLI, G., SIRONI, A. SARAVELLE, C., MUNARI, F., GALLI, M., GASPAR, G., COLIN, J. M., JOFELIN, J. L. (1985) J. High Res. Chromatogr. **8**: 771

36. LIPSKY, S. R., DUFFY, M. L. (1986a) J. High Res. Chromatogr. **9**: 376

37. LIPSKY, S. R., DUFFY, M. L. (1986b) J. High Res. Chromatogr. **9**: 725

38. STARK, T. J., LARSON, P. A., DANDENEAU, R. D. (1983) J. Chromatogr. **279**: 31

39. KOVATS, E. sz. (1958) Helv. Chim. Acta **41**: 1915

40. SHIBAMOTO, T., HARADA, K., YAMAGUCHI, K., AITOKU, A. (1980) J. Chromatogr. **194**: 277

41. GROB, K., GROB, G. (1983) Chromatographia. **17**: 481

42. KARLSEN, J., SIWON, H. (1975) J. Chromatogr. **110**: 187

43. SAEED, T., REDANT, G., SANDRA, P. (1979) J. High Res. Chromatogr. **2**: 75

44. WRIGHT, B. W., PEADEN, P. A., LEE, M. L., STARK, T. J. (1982) J. Chromatogr. **248**: 17

45. PERSINGER, H. E., SHANK, J. T. (1973) J. Chromatogr. Sci. **11**: 190

46. CONDER, J. R., FRUITWALA, N. A., SHINGARI, M. K. (1983) J. Chromatogr. **269**: 171

47. VERZELE, M., REDANT, G. R., VAN ROELENBOSCH, M., GODEFROOT, M., VERSTAPPE, M., SANDRA, P. (1981) Proc. 4th Int. Symp. on Capillary Chrom. (Kaiser, R. E., Ed.), Huethig Verlag, Heidelberg, p. 239

48. VERZELE, M., SANDRA, P. (1978) J. Chromatogr. **158**: 111

49. SANDRA, P., VERZELE, M., VERSTAPPE, M., VERZELE, J. (1979) J. High Res. Chromatogr. **2**: 288

50. HLAVAY, J., BARTHA, A., VIGH, G., GAZDAG, M. M., SZEPESI, G. (1981) J. Chromatogr. **204**: 59

51. DENIJS, R. C. M., DEZEEUW, J. (1982) J. High Res. Chromatogr. **5**: 501

52. TRAITLER, H., KOLAROVIC, L., SORIO, A. (1983) J. Chromatogr. **279**: 69

53. GROB, K. (1986) "Making and Manipulating Capillary Columns for Gas Chromatography", Huethig Verlag, Heidelberg, Basel, New York

54. BUIJTEN, J., BLOMBERG, L., MARKIDES, K., WANNMANN, T. (1983) J. Chromatogr. **268**: 387

55. BYSTRICKY, L. (1986) J. High. Res. Chromatogr. **9**: 240

56. ALLEN, R. R. (1966) Anal. Chem. **38**: 1287

57. WITHERS, M. K. (1972) J. Chromatogr. **66**: 249

58. HILTUNEN, R., RAISANEN, S. (1981) Planta Med. **41**: 174

59. GROB, K., Jr., GROB, K. (1977a) J. Chromatogr. **140**: 257

60. HAKEN, J. K. (1984) J. Chromatogr. **300**: 1

61. HAKEN, J. K. (1977) J. Chromatogr. **141**: 247

62. BLOMBERG, L. G., MARKIDES, K. E. (1985) J. High Res. Chromatogr. **8**: 632

63. BLOMBERG, L. (1982) J. High. Res. Chromatogr. **5**: 520

64. XU, B., VERMEULEN, N. P. E. (1988) J. Chromatogr. **445**: 1

65. BARTLE, K. D., WRIGHT, B. W., LEE, M. L. (1981) Chromatographia. **14**: 387

66. WRIGHT, B. W., PEADEN, P. A., LEE, M. L. (1982a) J. High Res. Chromatogr. **5**: 413

67. GOREN, S. L. (1961) J. Fluid Mech. **12**: 309

68. NOLL, W. (1968) "Chemistry and Technology of Silicones", Academic Press, New York

69. FOX, H. W., TAYLOR, P. W., ZISMAN, W. A., (1947) Ind. Eng. Chem. **39**: 1401

70. PEADEN, P. A., WRIGHT, B. W., LEE, M. L. (1982) Chromatographia. **15**: 335

71. STARK, T. J., LARSON, P. A. (1982) J. Chromatogr. Sci. **20**: 341

72. BLOMBERG, L., BUIJTEN, J., M ARKIDES, K., WANNMANN, T. (1982) J. Chromatogr. **239**: 51

73. BURNS, W. F., TINGLEY, D. T., EVANS, R. C. (1982) J. High Res. Chromatogr. **5**: 504

74. GROB, K., HABICH, A. (1983) J. High Res. Chromatogr. **6**: 34

75. GROB, K., GROB, G. (1982) J. High Res. Chromatogr. **5**: 349

76. JENNINGS, W. (1983) J. Chromatogr. Sci. **21**: 337

77. BEREZKIN, V. G., KOROLEV, A. A. (1985) Chromatographia. **20**: 482

78. BLOMBERG, L. (1984) J. High Res. Chromatogr. **7**: 232

79. WARWICK, E. L., PIERCE, O. R., POLMANTEER, K. E., SAAM, J. C. (1971) Rubber Chem. Tech. **52**: 437

80. TRESSL, R., ENGEL, K.-H. (1985) in "Progress in Flavor Research 1984" (Adda, P., Ed.), Elsevier, Amsterdam, p. 441

81. TRESSL, R., ENGEL, K.-H., ALBRECHT, W., BILLE-ABDULLAH, H. (1985) in "Characterization and Measurement of Flavor Compounds" (Bills, D. D.; Mussinan, C. J., Eds.), ACS Symposium Series 317, American Chemical Society. Washington, D. C., p. 43

82. CRAMERS, C. A., RIJKS, J. A., VAN ES, A. J., NOIJ, T. (1988) Proc. 9th Int. Symp. on Capillary Chromatogr., Huethig Verlag, Heidelberg, p. 1

83. KIRCHNER, J. C., MILLER, J. M. (1952) Ind. Eng. Chem. **44**: 318

84. KAISER, R., LAMPARSKY, D. (1976) Helv. Chim. Acta. **59**: 1797

85. TRESSL, R., FRIESE, L., FENDESACK, F., KOPPLER, H. (1978) J. Agric. Food Chem. **26**: 1422

86. IDSTEIN, H., HERRES, W., SCHREIER, P. (1984) J. Agric. Food Chem. **32**: 383

87. IDSTEIN, H., SCHREIER, P. (1985a) Lebensm. Wiss. u. Technol. **18**: 164

88. IDSTEIN, H., SCHREIER P. (1985b) J. Agric. Food Chem. **33**: 138

89. IDSTEIN, H., SCHREIER, P. (1985c) Phytochemistry. **24**: 2313

90. HUNTER, G. L. K., BROGEN, W. B., Jr. (1963) J. Org. Chem. **28**: 1679

91. WROLSTAD, R. E., JENNINGS, W. G. (1965) J. Chromatogr. **18**: 318

92. SNYDER, L. R. in "Chromatography" (1975) (Heftmann, E., Ed.), Van Nostrand Reinhold Company, New York, p. 54

93. KOVATS, E. SZ., KUGLER, E. (1963) Helv. Chim. Acta. **46**: 1480

94. VON RUDLOFF, E., COUCHMAN, F. M. (1964) Can. J. Chem. **42**: 1890

95. HEFENDEHL, F. W. (1970) Arch. Pharm. (Weinheim, Ger.). **303**: 345

96. KUBECZKA, K. H. (1973) Chromatographia **6**: 106

97. SCHEFFER, J. J. C., KOEDAM, A., BAERHEIM-SVENDSEN, A. (1976) Chromatographia. **9**: 425

98. MILES, D. H., MODY, N. V., MINYARD, J. P., HEDIN, P. A. (1975) Phytochemistry. **14**: 599

99. MAARSE, H., VAN OS, F. H. L. (1973) Flavor Ind. **4**: 477

100. BITTEUR, C. S. (1984) Analysis. **12**: 51

101. MORIN, P., CAUDE, M., RICHARD, H., ROSSET, R. (1985) Analysis. **13**: 196

102. TEITELBAUM, C. L. (1977) J. Agric. Food Chem. **25**: 466

103. JONES, B. B., CLARK, B. C., Jr., IACOBUCCI, G. A. (1979) J. Chromatogr. **168**: 575

104. YAMAGUCHI, K., MIHARA, S., AITOKU, A., SHIBAMOTO, T. (1979) in "Liquid Chromatographic Analysis of Food and Beverages" (Charalambous, G., Ed.), Vol. 2, Academic Press, New York, p. 303

105. KUBECZKA, K. H. (1981) in "Flavour '81" (Schreier, P., Ed.), 3rd Weurman Symp., de Gruyter, Berlin, p. 345

106. CHAMBLEE, T. S., CLARK, B. C., Jr., RADFORD, T., IACOBUCCI, G. A. (1985) J. Chromatogr. **330**: 141

107. MORIN. P., CAUDE, M., RICHARD, H., ROSSET, R. (1986) J. Chromatogr. **363**: 57

108. CLARK, B. J., Jr., CHAMBLEE, T. S., IACOBUCCI, G. A. (1987) J. Agric. Food Chem. **35**: 514

109. ETZWEILER, F. (1988) J. High Res. Chromatogr. **11**: 449

110. ROERAADE, J., BLOMBERG, S., PIETERSMA, H. D. J. (1986) J. Chromatogr. **356**: 271

111. GILLEN, D. G., SCANLON, J. T. (1972) J. Chromatogr. Sci. **10**: 729

112. DEMOLE, E., ENGGIST, P., OHLOFF, G. (1982) Helv. Chim. Acta. **65**: 1785

113. DEBRAUWERE, J., VERZELE, M. (1976) J. Chromatogr. Sci. **14**: 296

114. VON RUDLOFF, E. (1960) Can. J. Chem. **38**: 631

115. VERZELE, M. (1971) in "Preparative Gas Chromatography" (Zlatkis, A.; Pretorius, V., Eds.), Wiley (Interscience), New York, p. 230

116. CROTEAU, R., RONALD, R. C. (1983) J. Chromatogr. Libr. **22B**: 147

117. WILLIAMS, P. J., STRAUSS, C. R., WILSON, B., MASSY-WESTROPP, R. A. (1982) J. Chromatogr. **235**: 471

118. ENGEL, K.-H., TRESSL, R. (1983) J. Agric. Food Chem. **31**: 998

119. SALLES. C., ESSAIED, H., CHALIER. P., JALLAGEAS, J. C., CROUZET, J. (1988) in "Bioflavor '87" (Schreier, P., Ed.), Walter de Gruyter & Co., Berlin, p. 145

120. GUNATA, Y. Z., BAYONOVE, C. L., BAUMES, R. L., CORDONNIER, R. E. (1985) J. Chromatogr. **331**: 83

121. STRAUSS, C. R., GOOLEY, P. R., WILSON, B., WILLIAMS, P. J. (1987) J. Agric. Food Chem. **35**: 519

122. STRAUSS, C. R., WILSON, B., WILLIAMS, P. J. (1988) J. Agric. Food Chem. **36**: 569

123. NUNEZ, A. J., GONZALES, L. F., JANAK, J. (1984) J. Chromatogr. **300**: 127

124. DROZD, J., NOVAK, J. (1979) J. Chromatogr. **165**: 141

125. KOLB, B. (Ed.) (1980) "Applied Headspace Gas Chromatography", Heyden, London

126. CHARALAMBOUS, G. (Ed.) (1978) "Analysis of Food and Beverages, Headspace Techniques", Academic Press, New York

127. TAKEOKA, G., JENNINGS, W. (1984) J. Chromatogr. Sci. **22**: 177

128. MACKU, C., JENNINGS, W. G. (1987) J. Agric. Food Chem. **35**: 845

129. KALLIO, H., RINE, S., PANGBORN, R. M., JENNINGS, W. (1987) Food Chem. **24**: 287

130. KALLIO, H., LEINO, M., SALORINNE, L. (1988) Proc. 9th Int. Symp. on Capillary Chrom., Huethig Verlag, Heidelberg, p. 191

131. WITTKOWSKI, R., BALTES, W., TAKEOKA, G., JENNINGS, W. G. (1988) Lebensmittelchem. Gerichtl. Chem. **42**: 108

132. BALLSCHMITER, K., MAYER, P., CLASS, T. (1986) Z. Anal. Chem. **323**: 334

133. BARRIE, J. A. (1966) J. Polym. Sci., Part A–1. **4**: 3081

134. MACKU, C., KALLIO, H., TAKEOKA, G., FLATH, R. (1988) J. Chromatogr. Sci. **26**: 557

135. WYLIE, P. L. (1986) Chromatographia. **21**: 251

136. LEAHY, M. M., REINECCIUS, G. A. (1984) in "Analysis of Volatiles: Methods and Applications" (Schreier, P., Ed.), Walter de Gruyter, West Germany, p. 19

137. GROB, K., HABICH, A. (1985) J. Chromatogr. **321**: 45

138. BURGER, B. V., MUNRO, Z. (1986) J. Chromatogr. **370**: 449

139. BICCHI, D., D'AMATO, A., DAVID, F., SANDRA, P. (1988) Proc. 9th Int. Symp. on Capillary Chrom., Huethig Verlag, Heidelberg p. 214

140. BLOMBERG, S., ROERAADE, J. (1988) J. High Res. Chromatogr. **11**: 457

141. JENNINGS, W. G., SEVENANTS, M. R. (1964) J. Food Sci. **29**: 796

142. EVERS, W. J., HEINSOHN, H. H., Jr., MAYERS, B. J., SANDERSON, A. (1976) in "Phenolic, Sulfur and Nitrogen Compounds in Food" (Charalambous, G.; Katz, I., Eds.), ACS Symposium Series No. 26, American Chemical Society, Washington, D. C., p. 184

143. WITHYCOMBE, D. A., MUSSINAN, C. J. (1988) J. Food Sci. **53**: 658

144. BRUNKE, E. J., ROJAHN, W. (1984) Dragoco Rep. **31**: 67

145. CHRETIEN-BESSIERE, Y., GARNERO, J., BENEZET, L., PEYRON, L. (1967) Bull. Soc. Chim. France: 97

146. NAVES, Y. R. (1967) Bull. Soc. Chim. France: 3152

147. PETTUS, J. A., Jr., MOORE, R. E. (1970) Chem. Crommun.: 1093

148. MÜLLER, D. G., GASSMANN, G., MARNER, F.-J., BOLAND, W., JAENICKE, L. (1982) Science **218**: 1119

149. IDSTEIN, H., SCHREIER, P. (1983) in "Proc. Euro. Food Chem. II", (Ital. Soc. Food Sci., Ed.), Rome, p. 119

150. BERGER, R. G., DRAWERT, F., KOLLMANNSBERGER, H., NITZ, S., SCHAUFSTETTER, B. (1985a) J. Agric. Food Chem. **33**: 235

151. BERGER, R. G., DRAWERT, F., KOLLMANNSBERGER, H., NITZ, S. (1985b) J. Food. Sci. **50**: 1655

152. BUTTERY, R. G., LING, L. C., JULIANO, B. O. (1982a) Chem. Ind. (London): 958

153. SCHIEBERLE, P., GROSCH, W. (1987) Z. Lebensm. Unters. Forsch. **185**: 111

154. ENGEL, K.-H., FLATH, R. A., BUTTERY, R. G., MON, T. R., RAMMING, D. W., TERANISHI, R. (1988) J. Agric. Food Chem. **36**: 549

155. MAARSE, H., VISSHER, C. A. (1985) in "Volatile Compounds in Foods – Qualitative Data", Supplement 2, CIVO-TNO, Zeist

156. RE, L., MAURER, R., OHLOFF, G. (1973) Helv. Chim. Acta. **56**: 1882

157. ACREE, T. E., NISHIDA, R., FUKAMI, H. (1985) J. Agric. Food Chem. **33**: 425

158. FISCHER, K.-H., GROSCH, W. (1987) Lebensm. Wiss. u. Technol. **20**: 233

159. KEMP, T. R., KNAVEL, D. E., STOLTZ, L. P. (1972) Phytochemistry **11**: 3321

160. TAKEI, S., ONO, M. (1939) J. Agric. Chem. Soc. Japan. **15**: 193

161. LAMPARSKY, D., SCHUDEL, P. (1971) Tetrahedron Lett.: 3323

162. SUNDT, E., WILLHALM, B., CHAPPAZ, R., OHLOFF, G. (1971) Helv. Chim. Acta. **54**: 1801

163. KOLOR, M. G. (1983) J. Agric. Food Chem. **31**: 1125

164. BUTTERY, R. G., LING, L. C. (1973) J. Agric. Food Chem. **21**: 745

165. MURRAY, K. E., BANNISTER, P. E., BUTTERY, R. G. (1975) Chem. Ind. (London): 974

166. BUTTERY, R. G., GARIBALDI, J. A. (1976) J. Agric. Food Chem. **24**: 1246

167. DEMOLE, E., ENGGIST, P., SAUBERLI, U., STOLL, M., KOVATS, E. (1970) Helv. Chim. Acta. **53**: 541

168. NURSTEN, H. E., WOOLFE, M. L. (1972) J. Sci. Food Agric. **23**: 803

169. ACREE, T. E., BRAELL, P., BUTTS, R. M. (1981) J. Agric. Food Chem. **29**: 688

170. BUTTERY, R. G., TERANISHI, R., LING, L. C. (1988) Chem. Ind. (London): 238

171. TAKEOKA, G. R., FLATH, R. A., GUENTERT, M., JENNINGS, W. (1988) J. Agric. Food Chem. **36**: 553

172. TOULEMONDE, B., RICHARD, H. M. J. (1983) J. Agric. Food Chem. **31**: 365

173. TAKEOKA, G. R., GUENTERT, M., MACKU, C., JENNINGS, W. (1986) in "Biogeneration of Aromas" (Parliment, T. H.; Croteau, R., Eds.), ACS Symposium Series 317, American Chemical Society, Washington, D. C., p. 53

174. SCHREIER, P., DRAWERT, F., HEINDZE, J. (1981) Lebensm. Wiss. u. Technol. **14**: 150

175. BUTTERY, R. G., SEIFERT, R. M., LUNDIN, R. E., GUADAGNI, D. G., LING, L. C. (1969) Chem. Ind. (London): 490

176. RODIN, J. E., HIMEL, C. M., SILVERSTEIN, R. M., LEEPER, R. W., GORTNER, W. A. (1965) J. Food Sci. **30**: 280

177. HONKANEN, E., PYYSALO, T., HIRVI, T. (1980) Z. Lebensm. Unters. Forsch.: 180

178. RYDER, W. S. (1966) in "Advances Chem. Ser." Vol. 56, p. 88

179. GUADAGNI, D. G., BUTTERY, R. G., TURNBAUGH, J. G. (1972) J. Sci. Food Agric. **23**: 1435

180. CRONIN, D. A., WARD, M. K. (1971) J. Sci. Food Agric. **22**: 477

181. BUTTERY, R. G. (1981) in "Flavor Research – Recent Advances" (Teranishi, R., Flath, R. A., Sugisawa, H., Eds.), Marcel Dekker, Inc., New York, p. 175

182. BUTTERY, R. G., SEIFERT, R. M., LING, L. C., SODERSTROM, E. L., OGAWA, J. M., TURNBAUGH, J. G. (1982) J. Agric. Food Chem. **30**: 1208

183. MORITA, K., KOBAYASHI, S. (1966) Tetrahedron Lett.: 573

184. WADA, S., NAKATANI, H., MORITA, K. (1967) J. Food Sci. **32**: 559

185. HULTIN, H. O., PROCTOR, B. E. (1961) Food Technol. **15**: 440

186. SCHNIZ, H., SEIDEL, C. F. (1961) Helv. Chim. Acta. **44**: 278

187. SCHMIDLIN-MESZAROS, J. (1971) Alimentia **10**: 39

188. FLATH, R. A., BLACK, D. R., GUADAGNI, D. G., MCFADDEN, W. H., SCHULTZ, T. H. (1967) J. Agric. Food Chem. **15**: 29

189. SEIFERT, R. M., BUTTERY, R. G., GUADAGNI, D. G., BLACK, D. R., HARRIS, J. G. (1970) J. Agric. Food Chem. **18**: 246

190. BUTTERY, R. G., HADDON, W. F., SEIFERT, R. M., TURNBAUGH, J. G. (1984) J. Agric. Food Chem. **32**: 674

191. JENNINGS, W. G., CREVELING, R. K., HEINZ, D. E. (1964) J. Food Sci. **29**: 730

192. YAMAGUCHI, K., SHIBAMOTO, T. (1981) J. Agric. Food Chem. **29**: 366

II.4 Stereodifferentiation of Chiral Flavor and Aroma Compounds

K.-H. Engel

1 Introduction

Optical activity is a ubiquitous phenomenon in the chemistry of natural systems. Accordingly there are many chiral molecules among the naturally occurring flavor and aroma compounds. Comparable to the influence of the stereochemistry on the biological activities of pharmaceuticals [1] and pheromones [2], sensory properties of flavor and aroma compounds can also depend on their configurations. Many examples demonstrate this principle of enantioselectivity in odor perception [3, 4].

In the past, however, systematic investigations of the naturally occurring enantiomeric compositions of chiral flavor and aroma compounds have been lacking; this is mainly due to the fact that these volatiles are mostly contained as trace constituents and conventional chiroptical methods [5, 6] cannot be applied. NMR-spectroscopic methods, based on the use of chiral shift reagents or the investigation of diastereomeric derivatives, have been developed [7, 8]. Chromatographic techniques are alternatives requiring less sophisticated and less expensive instrumentation. High pressure liquid chromatography [9] and thin layer chromatography [10] can be employed to resolve optical isomers. For the analysis of chiral volatiles high resolution gas chromatography, a technique characterized by simplicity, sensitivity and accuracy, represents the method of the art.

Gas chromatographic resolution of chiral compounds can be achieved by two different approaches. *Indirect* methods are based on Pasteur's principle of converting enantiomers to diastereomeric derivatives by reaction with an optically pure reagent; however, the classical techniques employed to separate these diastereoisomers, e.g. crystallization or fractional distillation are replaced by the more efficient gas-liquid chromatography. *Direct* separations are due to differences in the association of enantiomers to an optically active stationary phase in the course of the chromatographic process. Comprehensive overviews covering the tremendous progress for both techniques in the last decades are available [11–20].

Investigations of the chirality of flavor and aroma compounds is a field of increasing importance to both basic research and food quality control. Some of the reviews concentrate on recent developments in the analysis of chiral volatiles [21, 3]. However, for someone intending to begin in this field the flood of information can be overwhelming and difficult to judge. Therefore, this contribution will not present another complete survey but select those data which are of special importance to the analysis of flavor and aroma compounds. The objective is to focus the available information in a way that it may help in the selection of the most appropriate method for a specific problem in the analysis of chiral volatiles.

The first part of the chapter will briefly summarize principles of both direct and indirect capillary GC separations of enantiomers and discuss their scopes and limits with regard to chiral flavor compounds. The second part will be devoted to applications of these techniques to the determination of naturally occurring enantiomeric compositions in various foods, the required basis for possible differentiation of "natural" and "nature-identical" flavor and aroma substances by means of chiral analysis.

2   Analytical Techniques

2.1   Indirect Methods

2.1.1   Derivatizing Reagents

Starting from the first reports in the early 1960's [22, 23] a large number of reagents suitable for formation of diastereomers which can be separated by means of gas chromatography have been employed. Derivatization strategies, aspects of the mechanisms of resolution and applications have been reviewed extensively by GIL-AV and NUROK [11]. In recent years the tremendous increase of separation efficiency in high resolution capillary gas chromatography led to an additional boost in this field. Four major reagents, currently used for chirality investigations of flavor and aroma compounds, have been selected to demonstrate advantages and limits of this technique. Their structures are shown in Figure II.4.–1.

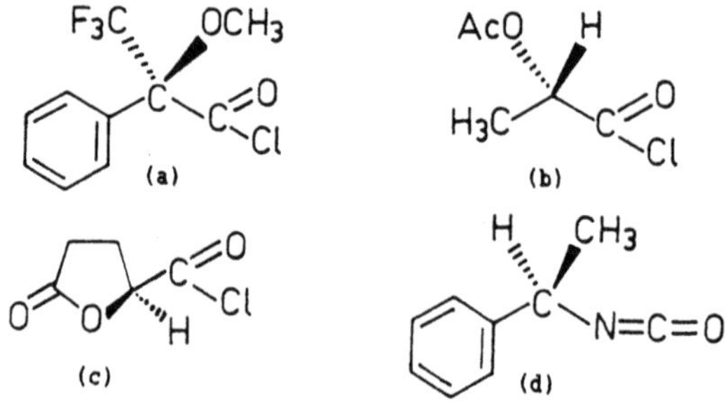

Fig. II.4.–1 Structures of major derivatizing reagents

    (a): (S)-α-methoxy-α-trifluoromethylphenylacetyl chloride
       (S)-MTPA-Cl

    (b): (S)-O-acetyllactyl chloride
       (S)-O-AL-Cl

    (c): (S)-tetrahydro-5-oxo-2-furancarboxylic acid chloride
       (S)-TOF

    (d): (R)-1-phenylethyl isocyanate
       (R)-PEIC

Table II.4.–1 Application of derivatizing reagents to chirality investigations of different classes of flavor and aroma compounds

Classes of compounds	Derivatizing reagents	References
Alkan-2-ols	(S)-O-AL-Cl	[23], [27], [29], [30]
	(R)-PEIC	[28], [31]
	(S)-MTPA-Cl	[27], [31], [33]
	(S)-TOF	[27]
3-Hydroxyacid esters	(S)-MTPA-Cl	[31], [32], [33]
	(R)-PEIC	[31]
	(S)-TOF	[34]
4-Hydroxyacid esters	(S)-MTPA-Cl	[33]
	(R)-PEIC	[31]
5-Hydroxyacid esters	(R)-PEIC	[31]
	(S)-TOF	[34]

(S)-(+)-α-methoxy-α-trifluoromethylphenylacetyl chloride ((S)-MTPA-Cl) was originally developed for assessment of optical purities by means of NMR-spectroscopy [24, 25]. The compound, known as Mosher's reagent, has become a versatile auxiliary for HPLC and GC separations. (S)-O-acetyllactyl chloride ((S)-O-AL-Cl) is one of the pioneer reagents used for the first separations of diastereomeric esters of secondary alcohols [23, 26]. (S)-tetrahydro-5-oxo-2-furancarboxylic acid chloride ((S)-TOF) is an example of a reagent which can be synthesized easily from the inexpensive naturally occurring (L)-glutamic acid [27], an interesting aspect for preparative applications. Not only diastereomeric esters but also carbamates, obtained after derivatization with (R)-1-phenylethyl isocyanate ((R)-PEIC), can be separated by capillary GC [28]. Table II.4.–1 presents an overview on applications of these derivatizing reagents to chirality investigations of flavor and aroma compounds.

2.1.2 Evaluation of Derivatizing Reagents

The gas chromatographic separation factors α (quotients of the retention times) of the derivatives obtained are the major criteria in evaluating the usefulness of a chiral compound as derivatizing reagent. Each reagent can be especially suitable for certain classes of compounds. Structural influences on the separation factors of diastereomers are demonstrated in Table II.4.–2. The position of the hydroxy group decisively influences the applicability of reagents to the investigation of chiral hydroxyacid esters.

Table II.4.–2   Structural influences on the separation factors of diastereomeric
derivatives of hydroxyacid esters

Compounds	Separation Factors (α)[a]		
	(S)-MTPA-Cl	(R)-PEIC	(S)-O-AL-Cl
Ethyl 2-hydroxyhexanoate	1.193	1.015	1.141
Ethyl 3-hydroxyhexanoate	1.065	1.031	1.082
Ethyl 5-hydroxyhexanoate	1	1.080	1.055

a = column DB-210 (J&W)

A major advantage of chirality investigations via diastereomeric derivatives is the fact that a broad spectrum of commonly available, achiral stationary phases can be used. Systematic investigations can provide the basis for selecting the combination of derivatizing reagent and capillary column which is the most appropriate for a certain class of chiral compounds [13, 27].

Among the variety of stationary phases employed for chirality investigations trifluoropropylmethylsilicone (commercially available as chemically bonded fused silica column, DB-210, J&W) turned out to be outstandingly suitable for the separation of diastereomeric derivatives. Resolution of all types of diastereoisomers listed in Table II.4.–1 can be achieved on this phase. The column exhibits high stability in the course of the analysis of complex mixtures; due to its intermediate polarity time consuming work up of derivatization mixtures, necessary if polar stationary phases, such as DB-Wax are used, is not required.

In addition to the separation factors, various other parameters, such as optical purity of the reagent, time required for the derivatization process, and volatility of the formed derivatives, have to be considered in choosing the appropriate auxiliary. They can seriously limit the usefulness of a reagent.

The availability of the reagent in optically pure form is an essential prerequisite. To a certain extent optical impurities can be considered by means of calculations [13]; accurate determinations of high enantiomeric excess, however, can only be carried out if the reagent is optically pure. (R)-1-phenylethyl isocyanate, for example, is a suitable reagent leading to excellent separation factors for various classes of compounds (Table II.4.–1); the optical purity of the commercially available compound, however, varies considerably and each batch of reagent should be checked before use.

To rule out errors, caused by enantiomeric discrimination in the course of the derivatization process, the conversion has to proceed quantitatively. Some reagents, e.g. (S)-O-Al-Cl, exhibit high reactivities resulting in reaction times of only a few minutes [29]. In cases of sterically hindered reactions, such as the derivatization of the tertiary alcohol linalool with (R)-PEIC [35] or the formation of dicarbamates from 1.4- and 1.5-diols [36], the use of a catalyst and prolonged reaction times are required. However, conditions leading to partial racemizations have to be avoided at any rate.

The volatility of diastereomeric derivatives can be a limiting factor in gas chromatographic analysis. (S)-O-acetyllactyl esters are rather volatile; on the other hand hydrogen had to be used as carrier gas to achieve elution and partial separation of di-[(R)-1-phenylethyl] carbamates derived from 1.4- and 1.5-diols on a DB 210 column without exceeding its temperature limit [36].

In most cases the elution order of diastereomeric derivatives remains constant within a homologous series. There are, however, significant exceptions, which demonstrate that optically pure reference compounds are needed for unambiguous determination of absolute configurations [29, 37].

2.1.3   Derivatization Strategies for γ- and δ-Lactones

If chiral compounds do not have reactive functional groups, investigations via diastereomeric derivatives can be a difficult task. Laborious reaction sequences had to be worked out to convert chiral monoterpene hydrocarbons to diastereomeric ketals which can be separated by capillary GC [38]. An important class of flavor and aroma compounds lacking reactive moieties are γ- and δ-lactones (4- and 5-alkanolides). Due to their pronounced sensory properties they play significant roles as flavor constituents in several foods [41]. Chirality investigations of γ- and δ-lactones are of particular interest and various efforts have been made to accomplish gas chromatograhic separation techniques.

Formation of diastereomeric ketals obtained by direct reaction with optically pure 2.3-butanediol has been described for δ-lactones [32, 40]. In general, however, a conversion of lactones to intermediates with reactive functional groups is required before diastereomeric derivatives can be formed. Table II.4.–3 summarizes derivatization strategies (combinations of intermediates and suitable derivatizing reagents) employed to obtain diastereoisomers, which can be separated by means of capillary GC.

Some of the procedures applied are relatively straightforward, e.g. the reduction of lactones to 1.4.- and 1.5-diols by means of lithiumaluminumhydride [34, 36, 42]. The applicability of others, such as the formation of hydroxyacid isopropylesters, is limited by reaction times unacceptable for routine analysis [44]. The major drawback of these procedures is the strong effect of lactone structure and chain length on the separation factors of diastereomeric derivatives. The results obtained for (S)-O-acetyllactyl esters of 1.4-diols and 4-hydroxyacid isopropylesters, respectively, [29, 42], are complementary (Fig. II.4.–2). Replacement of the acetyl moiety of (S)-O-AL-Cl by longer acyl side chains slightly improved the separations factors obtained for the higher lactone homologues [42]. However, a combination of both approaches would be needed to cover the complete spectrum of relevant lactones.

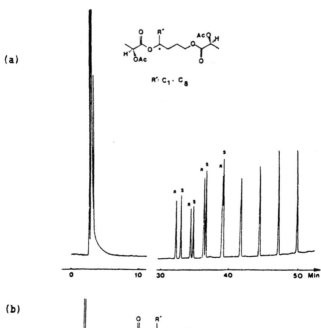

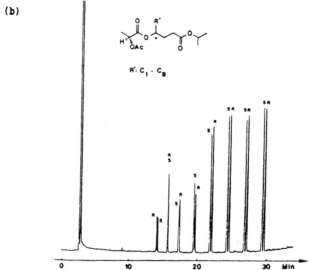

Fig. II.4.–2 Capillary gas chromatographic stereodifferentiation of γ-lactones via

(a): 1.4-diols as diesters of (S)-O-acetyllactic acid (DB-210, 30 m, N$_2$: 0.8 bar, 140 °C/ 2 °/min.)

(b): 4-hydroxyacid isopropylesters as (S)-O-acetyllactyl esters (DB-210, 30 m, N$_2$: 0.8 bar, 140 °C/2.5 °/min.) [29]

Table II.4.–3 Derivatization strategies for γ- and δ-lactones

Intermediates	Reagents	Ref.
Methyl and ethyl 4/5-hydroxyacid esters	(S)-MTPA-Cl	[32], [33]
	(R)-PEIC	[31]
Isopropyl 4/5-hydroxyacid esters	(S)-O-AL-Cl	[29]
	(S)-TOF	[39]
1.4/1.5-Diols	(S)-O-AL-Cl	[42]
	(S)-TOF	[34]
	(R)-PEIC	[36]
N-Butyl 4/5-hydroxycarboxamides	(R)-PEIC	[43]

The best results are obtained after conversion of lactones to N-butyl 4/5-hydroxycarbox-amides and subsequent derivatization with (R)-(+)-1-phenylethyl isocyanate [43]. This derivatization strategy leads to gas chromatographic separation factors high enough to investigate the chirality of the complete series of γ- (C_5-C_{12}) and δ-(C_6-C_{12})-lactones, those members of the class of lactones which are of special interest as constituents of fruits and vegetables (Fig. II.4.–3).

2.2   Direct Methods

2.2.1   Chiral Recognition via Hydrogen-Bonding

The first pioneering results in the area of direct gas chromatographic separations of enantiomers have been reported in 1966 by GIL-AV et al. [45, 46]. They made use of the chirality of amino acid building blocks to prepare optically active stationary phases for capillary columns. The separation of enantiomers is caused by the different stabilities of diastereomeric complexes, formed during the chromatographic process by reversible association between the chiral amino acid selectors in the stationary phase and the enantiomers.

Hydrogen bondings and dipole-dipole interactions play the most important roles. Modifications of the amino acid, peptide and ureide selectors extended the scope of this method from amino acids to a broad spectrum of chiral compounds [15].

The low temperature stability, however, remained a general drawback of these chiral stationary phases. This problem has been overcome by coupling the chiral amide moieties

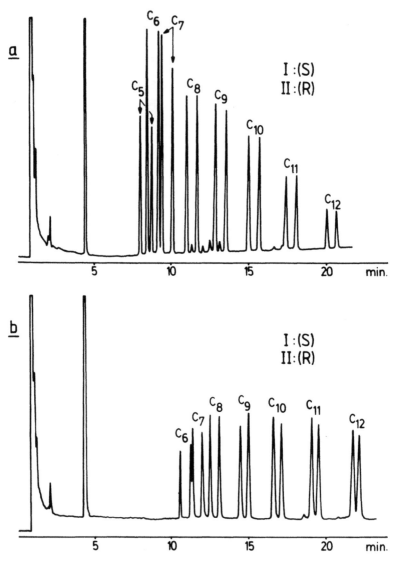

Fig. II.4.–3 Capillary gas chromatographic separation of 4- and 5-[(R)-1-
phenylethylcarbomoyloxy]-N-butylcarboxamides derived from γ-lactones
(Fig. **a**) and δ-lactones (Fig. **b**)

(DB-210, J&W, 30 m/0.32 mm i.d., film thickness 0.25 μm, hydrogen:
0.85 bar, 220 °C for 5 min. then progammed to 240 °C with a rate of
1 °/min.) [43]

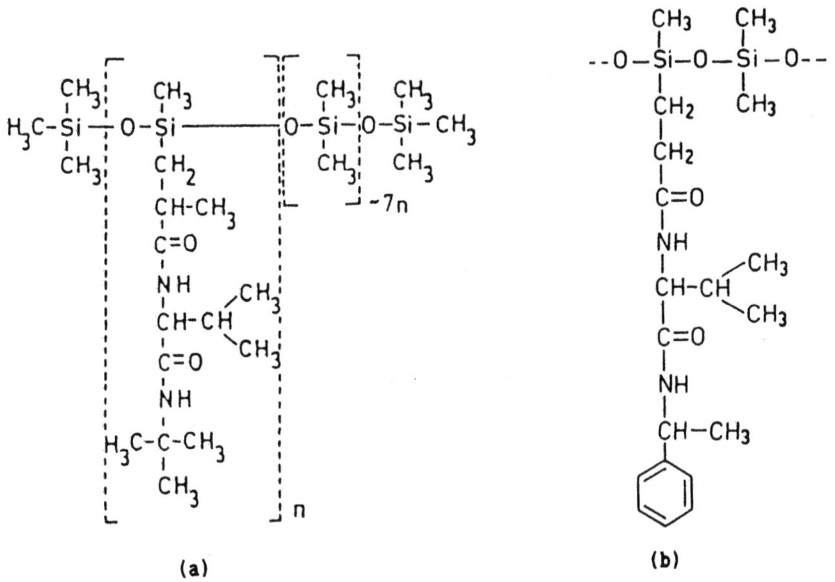

Fig. II.4.–4   Structures of **(a)** Chirasil-Val [47] and **(b)** XE 60-L-valine-(S)/(R)-1-phenylethylamide [49, 50]

with polysiloxanes. Figure II.4.–4 presents the structures of the two major polymeric chiral phases. Chirasil-Val has been introduced by BAYER et al. [47, 48]; they bonded L-valine-tert.-butyl-amide to a copolymer of dimethylsiloxane and (2-carboxypropyl)methylsiloxane. KÖNIG et al. [49, 50] modified XE-60-polysiloxane by connecting the polymeric skeleton with diasteromeric side chains consisting of (L)-valine-(R)- or (S)-1-phenylethylamide.

For most compounds chiral recognition by the stationary phase can only be achieved by formation of derivatives with additional C-O or CO-NH containing functional sites, thus increasing the molecular interactions between chiral selectors and enantiomers. Isocyanates [51] and phosgene [52] proved to be universal reagents for the derivatization of chiral compounds.

The formation of isopropyl urethanes made possible the separation of a broad spectrum of chiral alcohols, such as the homologous series of alkan-2-ols [53]. Further examples of chiral aroma compounds separated on chiral amide phases are octan-3-ol, terpinene-4-ol, isoborneol, α-bisabolol and α-ionol [54]. The resolution of the eight stereoisomers of p-menthan-3-ol could be achieved by means of multidimensional capillary GC, involving a preseparation of diastereomers on a non-chiral stationary phase and subsequent separation of enantiomers on the chiral column [51, 55]. Enantiomers of 2- and 3-hydroxycarboxylic acids could be separated after conversion to isopropylcarbomoyloxy-N-isopropylcarboxamides [51, 56] and tert.-butylcarbomoyloxy-N-tert.-butylcarboxamides [57], repectively. Isopropyl urethane derivatives were also used to separate α-hydroxy ketones, such as acetoin [58]. Volatile ketones, e.g. fenchone, camphor and menthone were converted to oxime derivatives by reaction with hydroxyl-ammonium chloride [59, 60]; the application of the procedure, however, is limited by the formation of syn- and anti-isomers.

Due to efficient deactivation of the inner column surface and the use of fused silica material enantiodifferentiation without previous derivatization could be achieved for certain classes of compounds [61, 62]. The four stereoisomeric sulfoxides, derived from cis-2-methyl-4-propyl-1.3-oxathiane can be separated on Chirasil-Val without derivatization [37, 63]. In general, however, prederivatizations are required; this limits the applicability of the technique to compounds with suitable functional groups (comparable to the indirect method). Low conversion rates, formation of by-products, and partial racemizations in the course of these derivatization steps can limit the accuracy of chirality investigations [64].

The high temperature stability and the broad spectrum of classes of chiral compounds which can be resolved are major advantages of chiral polysiloxane phases. Modifications of the chiral selectors may further enlarge the selectivity [65]. Immobilization techniques [66−68] will increase their stability and open the way for further applications, e.g. in capillary supercritical fluid chromatography [69, 70].

2.2.2 Chiral Recognition via Coordination

This technique employs organometallic chelate complexes derived from optically active monoterpene diketonates and transition metal ions as chiral selectors. Starting from the first separation of the enantiomers of an underivatized chiral olefin in 1977 [71] SCHURIG et al. developed a variety of chiral additives to stationary phases, thus establishing "complexation gas chromatography" as a valuable analytical tool with a wide scope of applications [16, 18, 21]. Some of the major "Chira-Metal" coordination compounds currently used are shown in Figure II.4.−5.

The major advantage of complexation GC is the fact that compounds capable of coordination to metal ions can be analyzed without any previous derivatization steps. Table II.4.−4 presents a selection of chiral aroma compounds which have been separated on Chira-Metal stationary phases. The separation of the four stereoisomers of 2-methyl-4-propyl-1.3-oxathiane, important flavor constituents of passion fruits is an impressive example [75] (Fig. II.4.−6).

Due to extremely long retention times and peak-tailings, complexation gas chromatography cannot be applied to direct separations of lactone enantiomers [21]. The detour via 1.4-diol acetonides (1.3-dioxepanes) and subsequent analysis on Chira-Metal phase (IV) can be considered as practical alternative for γ-lactones [77].

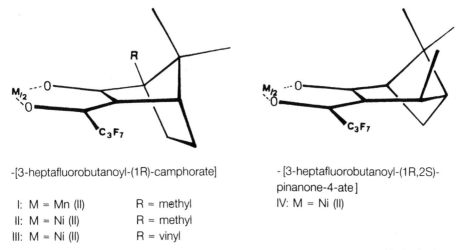

-[3-heptafluorobutanoyl-(1R)-camphorate]

- [3-heptafluorobutanoyl-(1R,2S)-
pinanone-4-ate]

I: M = Mn (II)   R = methyl   IV: M = Ni (II)
II: M = Ni (II)   R = methyl
III: M = Ni (II)   R = vinyl

Fig. II.4.– 5 Examples of Chira-Metal coordination compounds used as chiral selectors
in complexation gas chromatography

Most of the possible sources of error in determinations of optical purities by means of complexation GC can be eliminated by choosing the appropriate experimental conditions [20, 79]. However, the low temperature stability (120 °C) of Chira-Metal stationary phases is a decisive drawback and limits the applicability of this technique to highly volatile compounds. The investigation of complex mixtures, as usually obtained by extractions from natural products, can be difficult. A laborious work up procedure was required before the naturally occurring configuration of cis-2-methyl-4-propyl-1.3-oxathiane in yellow passion fruit could be determined by means of complexation gas chromatography [78].

The combination of various chiral monoterpene units with metal ions offers the possibility to create tailor-made Chira-Metal stationary phases for each specific separation problem. So far, however, no single stationary phase, covering most of the chemical classes of compounds, which are of interest in enantiomer analysis, is available. This narrow specificity of each stationary phase and the relatively high costs of purchase of the capillary columns may be the major reasons that complexation gas chromatography is not as widely used as one might expect from the potential scope of this elegant technique.

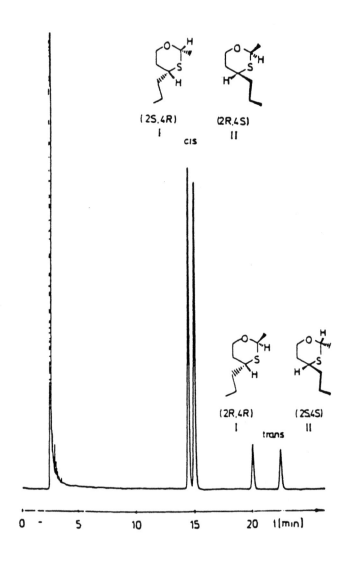

Fig. II.4.–6 Capillary gas chromatographic stereodifferentiation of cis/trans-2-methyl-4-
propyl-1.3-oxathiane on Ni(II)-bis-[3-(heptafluorobutanoyl)-(1R)-
camphorate], 0.125 M in OV-101, fused silica column, 25 m, 90 °C
isothermal [75, 78]

Table II.4.–4 Flavor and aroma compounds separated on Chira-Metal stationary phases

Compound	Sta. Phase[a] (Ref.)	Compound	Sta. Phase[a] (Ref.)
Monoterpene alcohols		**Esters**	
Menthol	II/[72]	Methyl 2-methylbutanoate	III/[21]
Neomenthol	II/[73]	Isobutyl 2-methylbutanoate	III/[21]
Isomenthol	II/[73]	Tert. butyl 2-methylbutanoate	III/[21]
Borneol	II/[73]	Isopropyl 2-methylpentanoate	III/[21]
Isoborneol	II/[73]	1-Octen-3-yl acetate	II/[37]
α-Fenchol	II/[73]	1-Octen-3-yl propanoate	II/[37]
Terpinene-4-ol	I/[74]	1-Octen-3-yl butanoate	II/[37]
Monoterpene ketones		**Cyclic ethers**	
Menthone	II/[72]	Cis-roseoxide	IV/[21]
Isomenthone	II/[72]	Trans-roseoxide	IV/[21]
Campher	IV/[73]	Neroloxide	IV/[21]
Acetals		**1.3-Dioxolanes**	
2-Methyl-4-propyl-1.3-Oxathianes	II/[75]	2-Alkyl-4-methyl-1.3-dioxolanes	II,IV/[76]

a = the number of the stationary phase corresponds to Figure II.4.–5

2.2.3 Chiral Recognition via Inclusion

The latest development in the field of direct gas chromatographic analysis of enantiomers is based on their inclusion in the cavities of chiral oligosaccharides. Cyclodextrins – cyclic oligomers of 6 (α-cyclodextrin), 7 (β-cyclodextrin) or 8 (γ-cyclodextrin) α-1.4 bonded glucose units – proved to be extraordinarily useful. Enantiomers of unsaturated hydrocarbons had been separated by using underivatized and permethylated β-cyclodextrin as stationary phases [80, 81]. The decisive breakthrough has been achieved by introduction of lipophilic side-arms; regioselective alkylation and acylation of cyclodextrins [82, 83] extended the scope of application to a broad spectrum of chiral compounds [84]. Commercially available modifications of cyclodextrins are shown in Figure II.4.–7.

Trifluoroacetylation and conversion to cyclic carbonates have been applied as prederivatization procedures for compounds with hydroxy- and amino-groups [82, 83]. However, direct separations without previous derivatizations have also been achieved; the outstand-

Lipodex A: $R^2 = R^3 = R^6$ = n-pentyl
Lipodex B: $R^2 = R^6$ = n-pentyl; R^3 = acetyl

Lipodex C: $R^2 = R^3 = R^6$ = n-pentyl
Lipodex D: $R^2 = R^6$ = n-pentyl; R^3 = acetyl

Fig. II.4.–7 Structures of commercially available chiral cyclodextrin stationary phases [84]

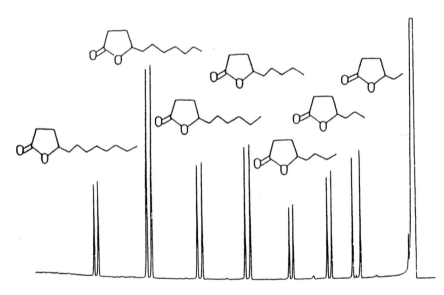

Fig. II.4.–8 Capillary gas chromatographic separation of γ-lactone enantiomers on Lipodex B (hexakis-(3-0-acetyl-2.6-di-O-pentyl-α-cyclodextrin); glass capillary column, 38 m; temp. program from 135 °C to 200 °C with a rate of 3 °/min [83]

ing example in the field of flavor and aroma compounds is the separation of γ-lactones using hexakis-(3-O-acetyl-2.6-di-O-pentyl)-α-cyclodextrin (Lipodex B) and heptakis (3-O-acetyl-2.6-di-O-pentyl)-β-cyclodextrin (Lipodex C) as stationary phases [83, 85] (Fig. II.4.–8). Other examples of flavor compounds separated are α-ionone, α-damascone, limonene, carvone and menthone [84].

The stationary phases are thermally very stable (> 200 °C). So far, however, only chemically non-bonded phases coated on glass capillary columns are available. Chirality investigations of compounds contained in complex mixtures have to be preceded by suitable purification steps. Liquid-solid chromatography, off-line coupling of HPLC and capillary GC, and compound transfer by means of multidimensional gas chromatography have been applied [85, 86]. Results concerning the chemical stability of the columns in the course of routine analysis of complex mixtures will have to be awaited. Suitable immobilization techniques and the use of fused silica material should further enlarge the applicability of this promising new technique.

2.3 Conclusions

Gas chromatographic separations of diastereomeric derivatives are useful tools for chirality investigations of flavor and aroma compounds, if various shortcomings, which can limit their accuracy, are considered carefully. The method is characterized by broad applicability; the fact that commonly used capillary columns can be employed makes it an especially useful alternative for laboratories with standard GC equipment (and low budget).

Undoubtedly direct separation of enantiomers on optically active stationary phases is the more elegant technique; especially if no prederivatization procedures are required. At present, their use is limited by high costs of commercially available columns on one hand and narrow specificity and a lack of results concerning their stability in the course of routine and analysis on the other hand. In the future, however, investigations on optically active stationary phases will definitely dominate the field of chirality investigations of flavor and aroma compounds.

3 Applications

The described gas chromatographic methods can be used to determine configurations of chiral constituents in natural systems, even if they are present only at trace levels. This information is extremely valuable for both fundamental research and practical applications in flavor production and food quality control.

To meet the increasing demand for "natural" flavor, basic knowledge about the pathways involved in the biogenesis of volatiles is needed. For chiral compounds the configuration offers an additional, valuable criterion to trace back biogenetical routes. Current strategies for the investigations of chiral flavor and aroma compounds are based on different approaches: (a) determination of naturally occurring configurations in plant or microbial systems, (b) studies on the stereochemical course of the metabolization of potential precursors, (c) isolation of enzymes involved and characterization of their enantioselectivity, and (d) use of commercially available enzymes or microorganisms as model systems to simulate "in vivo-conditions". This combination of methods revealed some new aspects of the biosynthesis of chiral flavor and aroma compounds [34, 87, 88].

This chapter will concentrate on determinations of naturally occurring configurations of chiral volatiles in plant systems and the possible application of these data to food quality control. The European legislation for the regulation of flavoring substances used in foods distinguishes between "natural" and "nature-identical" flavor and aroma compounds. "Natural" flavors must be isolated from natural sources by means of physical methods (e.g. distillation, extraction) or may be obtained by means of fermentative processes starting from natural material. "Nature-identical" flavoring substances can be produced via chemical synthesis but they have to be chemically identical to the naturally occurring compounds.

It can be difficult to determine the category to which an aroma constituent belongs. For chiral compounds their configuration should offer an additional criterion to differentiate between "natural" and "nature-identical" flavor substances. Generally, chemically synthesized compounds are racemic, unless (expensive) stereospecific catalysts are used. On the other hand, enzyme-catalyzed reactions are known to proceed with a high degree of enantioselectivity; therefore, optically active compounds are expected in natural systems.

The application of this concept, however, has been seriously complicated by results of investigations of naturally occurring configurations. They revealed that chiral flavor and aroma constituents are not always present in optically pure form but in some cases as mixtures of enantiomers. Detailed knowledge of the limits between which naturally occurring optical purities of a compound can vary is an essential premise before chirality investigations can be applied to determine the "natural" character of a flavoring substance. These data are rather scattered in literature; this chapter will present a summarizing compilation for selected classes of compounds.

3.1 Chiral Alcohols

Mushrooms contain the unsaturated chiral C_8-alcohol 1-octen-3-ol as aroma contributing constituent [89–91]. It had been known from polarimetric studies that the levorotatory enantiomer is naturally occurring in mushrooms [89]; its exact enantiomeric purity could be determined by means of capillary GC investigation of diastereomeric derivatives (Tab. II.4.–5).

Table II.4.–5 Naturally occurring enantiomeric compositions of 1-octen-3-ol

Natural Systems	1-Octen-3-ol		
	% (R)	% (S)	Ref.
Agaricus campestris	97	3	[30]
Psalliota bispora	93	7	[92]
Cantharellus cibarius	89	11	[30]
Red clover (Trifolium pratense L.)	100	–	[94]
Subterranean clover (Trifolium subterraneum)	100	–	[94]
Alfalfa (Medicago sativa)	99.5	0.5	[94]

High amounts of 1-octen-3-ol could also be isolated by means of simultaneous distillation-extraction from some forage legumes, such as clover and alfalfa [93]. These plants might be used as alternative natural sources for 1-octen-3-ol; chirality investigations by capillary GC revealed that they contain the same enantiomer as mushrooms (Table II.4.–5). This indicates that the enzyme-catalyzed stereospecific cleavage of the 10-hydroxyperoxide isomer of linoleic acid, elucidated as the decisive step in the formation of 1-octen-3-ol in mushrooms [92] might be a general principle in the biogeneration of this chiral alcohol.

Sensory evaluation of purified enantiomers demonstrated that only (R)-(–)-1-octen-3-ol is responsible for the fruity, mushroom-like flavor, whereas the (S)-antipode exhibits a moldy, grassy note, reminiscent of mushrooms, but less intensive [95]. A racemic mixture, however, is still potent enough to be used as flavoring substance in mushroom products. Due to the uniformly high optical purity of 1-octen-3-ol in natural systems the investigation of the configuration can be used to detect such adulterations of mushroom-based products by addition of chemically synthesized 1-octen-3-ol.

Odd-numbered alkan-2-ols and their esters are contained as volatile constituents in many fruits. The (S)-enantiomer of heptan-2-ol has been identified as major component in blackberries [34]. In contrast to 1-octen-3-ol, however, the optical purity of alkan-2-ols in plants cannot be generalized; there are other natural systems, such as passion fruits or bananas, where alkan-2-ols are not present in optically pure form but as mixtures of enantiomers (Tab. II.4.–6). These ratios vary depending on the chain lengths of the alcohols. This phenomenon is even more pronounced in corn [34]. As shown in Table II.4.–7 heptan 2-ol is mainly present as (R)-enantiomer; with increasing chain length the proportion of the (S)-antipode increases. A similar distribution of enantiomers has been found in coconuts [34].

Table II.4.–6 Naturally occurring configurations of alkan-2-ols and alkan-2-yl esters in passion fruits [33] and bananas [30, 96]

| | Passion Fruit | | | | Banana | |
| | yellow | | purple | | | |
	(R)	(S)	(R)	(S)	(R)	(S)
Pentan-2-ol	33 %	67 %	_b		_b	
Heptan-2-ol	14 %	86 %	92 %	8 %	39 %	61 %
(Z)-4-Hepten-2-ol	_a		_a		26 %	74 %
2-Heptyl esters	_a		100 %	–	–	100 %

a = not contained, b = not determined

Table II.4.–7 Enantiomeric ratios of secondary alcohols in corn varieties [34]

| Secondary alcohols | Bonanza | | Golden Bantam | | Golden Jubilee | |
	(S)	(R)	(S)	(R)	(S)	(R)
Heptan-2-ol	26 %	74 %	25 %	75 %	13 %	87 %
(Z)-4-Hepten-2-ol	24 %	76 %	11 %	89 %	13 %	87 %
Octan-2-ol	28 %	72 %	28 %	72 %	11 %	89 %
Nonan-2-ol	48 %	52 %	35 %	65 %	27 %	73 %
Decan-2-ol	63 %	37 %	43 %	57 %	24 %	76 %
Undecan-2-ol	77 %	23 %	98 %	2 %	41 %	59 %

The biogenesis of methyl ketones can be explained by modified β-oxidation of fatty acids [33]. Reductions of methyl ketone precursors by xerophilic fungi have been used as model systems to study the stereospecificity of the last step of the formation of secondary alcohols. The obtained enantiomeric ratios indicate the presence of at least two chain length specific oxidoreductases with different enantioselectivities [34].

In contrast to the free alcohols, alkan-2-yl esters in passion fruits and bananas are present in optically pure form (Tab. II.4.–6). The enzyme catalyzed esterification of secondary alcohols in fruits is characterized by high enantioselectivity. The presence of opposite enantiomers of heptan-2-ol in the two passion fruit varieties and the exclusive esterification of the (R)-antipode to (R)-alkan-2-yl esters in the purple fruits are a basis for a differentiation of the two varieties [33]. Bananas, on the other hand, contain the (S)-configurated esters [30]; these (S)-enantiomers exhibit the typical fruity, estery notes [97]. The enzymes involved in the strictly enantioselective esterification of secondary alcohols in fruits are not known as yet. Ester syntheses catalyzed by commercially available lipases in organic solvent have been investigated as model systems to get more insight into this process [98, 99].

3.2 Hydroxyacid Esters

Hydroxy- and acetoxyacid esters are typical constituents of various tropical fruits. Ethyl 3-hydroxyacid esters have been identified in passion fruit and mango [100, 101]; the corresponding methyl 3-, 4-, and 5-acetoxyacid esters are prominent constituents of pineapples [102, 103]. Enantiomeric ratios determined in various fruits by capillary GC separations of MTPA- and PEIC derivatives are listed in Tables II.4.−8 and II.4.−9.

Table II.4.−8 Enantiomeric compositions of 3-hydroxyacid esters in tropical fruits [33]

	Ethyl 3-hydroxybutanoate		Ethyl 3-hydroxyhexanoate	
	(R)	(S)	(R)	(S)
Yellow passion fruit	18 %	82 %		
Purple passion fruit	69 %	31 %		
Mango	78 %	22 %		
Purple passion fruit			85 %	15 %

Table II.4.−9 Enantiomeric compositions of hydroxy- and acetoxyacid esters in pineapple [31, 33]

Pineapple	(R)	(S)
Methyl 3-hydroxyhexanoate	9 %	91 %
Methyl 3-acetoxyhexanoate	7 %	93 %
Methyl 5-acetoxyhexanoate	21 %	79 %
Methyl 5-acetoxyoctanoate	47 %	53 %

The almost inverse enantiomeric ratios of ethyl 3-hydroxybutanoate and ethyl 3-hydroxy-hexanoate in purple passion fruits are similar to those obtained by reduction of the corresponding oxoprecursors by baker's yeast. The combination of enzyme purification steps and gas chromatographic chirality investigations revealed that two enzymes with different enantioselectivities and different substrate specificities, the so-called (R)- and (S)-enzymes, compete for the substrate in the course of the reduction of 3-oxoacid esters by baker's yeast [104]. Incubation of pineapple tissue with oxoacid esters revealed that the optical purities of the hydroxyacid esters formed depend on the chain lengths of the substrates [88]. This indicates a competition of oxidoreductases in plant systems comparable to baker's yeast.

If chiral compounds are present in natural systems as mixtures of enantiomers rather than in optically pure form, the constancy of such enantiomeric ratios is an essential premise for further use of this criterion in the evaluation of the "natural" origin of a flavoring substance. In pineapples subjected to postharvest ripening at room temperature over a period of five days the concentrations of chiral hydroxy- and acetoxyacid esters increased considerably, their enantiomeric compositions, however, remained nearly unchanged [88], (Fig. II.4–9).

3.3 γ- and δ-Lactones

Chirality investigations of hydroxy- and acetoxyacid esters may be considered as esoteric fundamental research; investigations of the related γ- and δ-lactones, however, are of greatest practical importance. These compounds contribute significantly to the aroma of many fruits; coconuts, peaches, apricots and nectarines are prominent examples [41, 108–110]. Due to their flavor and aroma qualities they are widely used as flavoring materials. The tremendous efforts made to establish capillary gas chromatographic methods for investigations of their optical purities (described in the first part of this chapter) reflect the significance of this class of compounds. Applications of these procedures to lactones isolated from natural systems revealed a high diversity of naturally occurring configurations; for some lactones ratios of enantiomers identified in various foods are summarized in Table II.4.–10. The results are rather complex: there are lactones occurring in optically pure form, some exhibit a significant excess of one enantiomer and others are present as almost racemic mixture.

Investigations of different oak woods revealed the presence of varying ratios of cis/trans isomers of 3-methyl-4-octanolide (the so-called "quercus" or "whisky" lactone); their configurations, however, are always 3S,4R (trans) and 3S,4S (cis), respectively [85, 107]. γ-Decalactone and γ-dodecalactone are contained as optically pure (R)-enantiomers in strawberries [44, 85].

In other fruits, such as mangos, peaches and nectarines, only an excess of the (R)-enantiomer was determined for γ-decalactone. δ-Octalactone is mainly present as (R)-enantio-

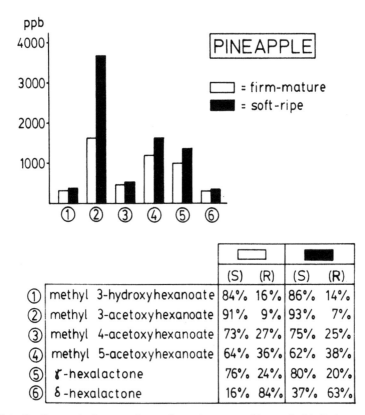

		(S)	(R)	(S)	(R)

		(S)	(R)	(S)	(R)
①	methyl 3-hydroxyhexanoate	84%	16%	86%	14%
②	methyl 3-acetoxyhexanoate	91%	9%	93%	7%
③	methyl 4-acetoxyhexanoate	73%	27%	75%	25%
④	methyl 5-acetoxyhexanoate	64%	36%	62%	38%
⑤	γ-hexalactone	76%	24%	80%	20%
⑥	δ-hexalactone	16%	84%	37%	63%

Fig. II.4.–9 Concentrations and enantiomeric compositions of chiral pineapple constituents at different stages of ripeness

mer in coconut, in pineapple, however, it is contained as almost racemic mixture. Investigations of nectarines demonstrated that the enantiomeric compositions of lactones remained nearly unchanged during the maturation period [34].

Mixtures of lactone enantiomers in natural systems may be rationalized by the complexity of their biogenesis. Addition of labelled precursors revealed that three independent pathways, (a) reduction of 5-oxoprecursors, (b) chain elongation of 3-hydroxyhexanoate, and (c) hydration of (Z)-4-octenoic acid are involved in the biosynthesis of δ-octalactone in pineapple [88]. If different enzymes competing for a common substrate or different pathways leading to the same product exhibit inverse enantioselectivities ratios of enantiomers rather than optically pure compounds can finally be obtained.

Table II.4.–10 Naturally occurring enantiomeric compositions of γ- and δ-lactones

Fruit	γ-Hexalactone		Ref.
	(R)	(S)	
Pineapple	85 %	15 %	[31]
	24 %	76 %	[88]
	20 %	80 %	[88]
Mango	45 %	55 %	[105]
Nectarine	86 %	14 %	[34]

Fruit	δ-Octalactone		Ref.
	(R)	(S)	
Pineapple	56 %	44 %	[31]
	51 %	49 %	[88]
	53 %	47 %	[88]
Coconut	92 %	8 %	[32]
	86 %	14 %	[30]

Fruit	γ-Decalactone		Ref.
	(R)	(S)	
Mango	72 %	28 %	[105]
Strawberry	100 %	–	[44]
			[85]
Peach	89 %	11 %	[106]
Nectarine	90 %	10 %	[34]

Fruit	δ-Decalactone		Ref.
	(R)	(S)	
Mango	89 %	11 %	[105]
Coconut	85 %	15 %	[32]
	71 %	29 %	[30]
Nectarine	95 %	5 %	[34]

3.4 Conclusions

If chiral aroma compounds, such as 1-octen-3-ol, are generally present in natural systems in optically pure form or at least with high enantiomeric excess, determinations of their configurations can be applied to detect the use of chemically synthesized, racemic flavoring substances. If chiral constituents occur as mixtures of enantiomers the following aspects have to be considered before conclusions concerning the natural origin of flavoring material can be drawn:

(a) What are the possible ranges of enantiomeric ratios? Examples presented in this chapter demonstrate that optical purities of chiral volatiles can be rather independent from the stage of maturity or ripeness. This has to be verified for other classes of compounds and other plant systems. Possible variations of optical purities depending on origin, variety or technological processing have to be determined.

(b) If (nearly) racemic mixtures of chiral compounds occur naturally, e.g. δ-octalactone in pineapple, the proof of an adulteration is even more difficult. A natural system containing such racemic mixtures can always be claimed as the source which had been used to isolate the flavoring substance. In such cases qualitative and quantitative data concerning the aroma profiles are additionally required.

(c) Biotechnological procedures leading to optically pure enantiomers are of increasing importance in the industrial production of chiral flavoring compounds. As long as fermentations starting from natural sources are applied the products obtained are to be considered as "natural"; biotransformations of chemically synthesized intermediates, however, lead to "nature-identical" compounds. This material might be used to adjust the optical purity of flavoring compounds to the "correct", naturally occurring enantiomeric ratios; the proof of such an adulteration would be difficult.

References

1. WAINER, I. W., DRAYER, D. E. (Eds.) (1988) "Drug Stereochemistry. Analytical Methods and Pharmacology", Marcel Dekker, New York, Basel

2. MORI, K. (1984) in "Techniques in Pheromone Research" (Hummel, H. E., Miller, T. A., Eds.), Springer Verlag, New York, Berlin, Heidelberg, Tokyo, p. 323

3. MOSANDL, A. (1988) Food Rev. Int. **4**: 1

4. PICKENHAGEN, W. (1989) in "Flavor Chemistry – Trends and Developments" (Teranishi, R., Buttery, R. G., Shahidi, F., Eds.), ACS Symposium Series 388, American Chemical Society, Washington, D. C., p. 151

5. SCHURIG, V. (1985) Kontakte (Darmstadt) **(1)**: 54

6. PURDIE, N., SWALLOWS, K. A. (1989) Anal. Chem. **61**: 77A

7. PLUMMER, E. L., STEWART, T. E., BYRNE, K., PEARCE, G. T., SILVERSTEIN, R. M. (1976) J. Chem. Ecol., **2**: 307

8. SCHURIG, V. (1985) Kontakte (Darmstadt) **(2)**: 22

9. ARMSTRONG, D. W. (1987) Anal. Chem. **59**: 84A

10. MACK, M., HAUCK, H. E. (1988) Chromatographia **26**: 197

11. GIL-AV, E., NUROK, D. (1974) Adv. Chromatogr. **10**: 99

12. LOCHMÜLLER, C. H., SOUTER, R. W. (1975) J. Chromatogr. **113**: 283

13. LIU, H., KU, W. W. (1981) Anal. Chem. **53**: 2180

14. KÖNIG, W. A. (1982) J. High Res. Chromatogr. **5**: 588

15. LIU, R. H., KU, W. W. (1983) J. Chromatogr. **271**: 309

16. SCHURIG, V. (1983) in "Asymmetric Synthesis" (Morrison, J. D., Ed.), Vol. I, Academic Press, New York, p. 59

17. SCHURIG, V. (1984) Angew. Chem. **96**: 733

18. SCHURIG, V. (1986) Kontakte (Darmstadt) No. **1**: 3

19. KÖNIG, W. A. (1987) "The Practice of Enantiomer Separation by Capillary Gas Chromatography", Huethig, Heidelberg, Basel, New York

20. SCHURIG, V. (1988) J. Chromatogr. **441**: 135

21. SCHURIG, V. (1988) in "Bioflavour '87" (Schreier, P., Ed.), Walter de Gruyter, Berlin, New York, p. 35

22. CASANOVA, J., COREY, E. J. (1961) Chem. Ind. (London) 1664

23. GIL-AV, E., NUROK, D. (1962) Proc. Chem. 146

24. DALE, J. A., HULL, D. L., MOSHER, H. S. (1969) J. Org. Chem. **34**: 2543

25. DALE, J. A., MOSHER, H. S. (1973) J. Am. Chem. Soc. **95**: 512

26. ROSE, H. C., STERN, R. L., KARGER, B. L. (1966) Anal. Chem. **38**: 469

27. DOOLITTLE, R. E., HEATH, R. R. (1984) J. Org. Chem. **49**: 5041

28. PEREIRA, W., BACON, V. A., PATTON, W., HALPERN, B. (1970) Anal. Lett. **3**: 23

29. MOSANDL, A., GESSNER, M., GÜNTHER, C., DEGER, W., SINGER, G. (1987) J. High Res. Chromatogr. **10**: 67

30. GESSNER, M., DEGER, W., MOSANDL, A. (1988) Z. Lebensm. Unters. Forsch. **186**: 417

31. TRESSL, R., ENGEL, K.-H., ALBRECHT, W., BILLE-ABDULLAH, H. (1985) in "Characterization and Measurement of Flavor Compounds" (Bills, D. D., Mussinan, C. J., Eds.), ACS Symposium Series 289, American Chemical Society, Washington, D. C., p. 43

32. TRESSL, R., ENGEL, K. H. (1984) in "Analysis of Volatiles" (Schreier, P., Ed.), Walter de Gruyter, Berlin, New York, p. 323

33. TRESSL, R., ENGEL, K. H. (1985) in "Progress in Flavor Research 1984" (Adda J., Ed.), Elsevier Science Publishers, Amsterdam, p. 441

34. ENGEL, K.-H. (1988) in "Bioflavour '87" (Schreier, P., Ed.), Walter de Gruyter, Berlin, New York, p. 75

35. RUDMANN, A. A., ALDRICH, J. R. (1987) J. Chromatogr. **407**: 324

36. ENGEL, K.-H., FLATH, R. A., ALBRECHT, W., TRESSL, R. (1989) J. Chromatogr. **479**: 176

37. DEGER, W., GESSNER, M., HEUSINGER, G., SINGER, G., MOSANDL, A. (1986) J. Chromatogr. **366**: 385

38. SATTERWHITE, D. M., CROTEAU, R. B. (1987) J. Chromatogr. **407**: 243

39. MOSANDL, A., GESSNER, M. (1988) Z. Lebensm. Unters. Forsch. **187**: 40

40. SAUCY, G., BORER, R., TRULLINGER, D. P., JONES, J. B., LOCK, K. P. (1977) J. Org. Chem. **42**: 3206

41. MAGA, J. A. (1976) Crit. Rev. Food Sci. Nutr., 1

42. DEGER, W., GESSNER, M., GÜNTHER, C., SINGER, G., MOSANDL, A. (1988) J. Agric. Food Chem. **36**: 1260

43. ENGEL, K.-H., ALBRECHT, W., HEIDLAS, J. (1990) J. Agric. Food Chem. **38**: 244

44. KRAMMER, G., FRÖHLICH, O., SCHREIER, P. (1988) in "Bioflavour '87" (Schreier, P., Ed.), Walter de Gruyter, Berlin, New York, p. 89

45. GIL-AV, E., FEIBUSH, B., CHARLES-SIGLER, R. (1966) Tetrahedron Lett. 1009

46. GIL-AV, E., FEIBUSH, CHARLES-SIGLER, R. (1967) in Littlewood, A. B. (Ed.), "Gas Chromatography 1966", Institute of Petroleum, London, p. 227

47. FRANK, H., NICHOLSON, G. J., BAYER, E. (1977) J. Chromatogr. Sci. **15**: 174

48. FRANK, H., NICHOLSON, G. J., BAYER, E. (1978) Angew. Chem. **90**: 396; Angew. Chem. Int. Ed. Engl. **17**: 363

49. KÖNIG, W. A., SIEVERS, S., BENECKE, I. (1981) in "Proc. 4th Int. Symp. on Capillary Chrom." (Kaiser, R. E., Ed.), Institute for Chromatography, Bad Dürkheim, and Huethig Publishers, Heidelberg, p. 703

50. KÖNIG, W. A., BENECKE, I., SIEVERS, S. (1981) J. Chromatogr. **217**: 71

51. BENECKE, I., KÖNIG, W. A. (1982) Angew. Chem. **94**: 709; Angew. Chem. Suppl. 1605

52. KÖNIG, W. A., STEINBACH, E., ERNST, K. (1984) Angew. Chem. **96**: 516

53. KÖNIG, W. A., FRANCKE, W., BENECKE, I. (1982) J. Chromatogr. **239**: 227

54. KÖNIG, W. A. (1984) in "Analysis of Volatiles" (Schreier, P., Ed.), Walter de Gruyter, Berlin, New York, p. 77

55. SCHOMBURG, G., HUSMANN, H., HÜBIGER, E., KÖNIG, W. A. (1984) J. High Res. Chromatogr. **7**: 404

56. FRANK, H., GERHARDT, J., NICHOLSON, G. J., BAYER, E. (1983) J. Chromatogr. **270**: 159

57. KÖNIG, W. A., BENECKE, J., LUCHT, N., SCHMIDT, E., SCHULZE, J., SIEVERS, S. (1983) J. Chromatogr. **279**: 555

58. RIECK, M., HAGEN, M., LUTZ, S., KÖNIG, W. A. (1988) J. Chromatogr. **439**: 301

59. KÖNIG, W. A., BENECKE, I., ERNST, K. (1982) J. Chromatogr. **253**: 267

60. KONIG, W. A., SCHMIDT, E., KREBBER, R. (1984) Chromatographia **18**: 698

61. KOPPENHÖFER, B., ALLMENDINGER, H., NICHOLSON, G. J., BAYER, E. (1983) J. Chromatogr. **260**: 63

62. KOPPENHÖFER, B., ALLMENDINGER, H., NICHOLSON, G,. (1985) Angew. Chem. **97**: 46

63. SINGER, G., HEUSINGER, G., MOSANDL, A., BURSCHKA, C. (1987) Liebigs Ann. Chem. 451

64. KOPPENHÖFER, B., ALLMENDINGER, H. (1987) Fresenius Z. Anal. Chem. **326**: 434

65. FRANK, H. (1988) J. High Res. Chromatogr. **11**: 787

66. BENECKE, I., SCHOMBURG, G. (1985) J. High Res. Chromatogr. **8**: 191

67. SCHOMBURG, G., BENECKE, I., SEVERIN, G. (1985) J. High Res. Chromatogr. **8**: 991

68. LAI, G., NICHOLSON, G., BAYER, E. (1988) Chromatographia **26**: 229

69. BRADSHAW, J. S., AGGARWAL, S. K., ROUSE, C. A., TARBET, B. J., MARKIDES, K. E., LEE, M. L. (1987) J. Chromatogr. **405**: 169

70. RUFFING, F.-J., LUX, J. A., ROEDER, W., SCHOMBURG, G. (1988) Chromatographia **26**: 19

71. SCHURIG, V. (1977) Angew. Chem. **89**: 113 Angew. Chem. Inter. Ed. Engl. **16**: 110

72. SCHURIG, V., WEBER, R. (1983) Angew. Chem. **95**: 797

73. SCHURIG, V., WEBER, R. (1983) Angew. Chem. Suppl. 1130

74. SCHURIG, V., LEYRER, U., KOHNLE, U. (1985) Naturwissensch. **72**: 211

75. MOSANDL, A., HEUSINGER, G., WISTUBA, D., SCHURIG, V. (1984) Z. Lebensm. Unters. Forsch. **179**: 385

76. MOSANDL, A., HANGENAUER-HENER, U. (1988) J. High Res. Chromatogr. **11**: 744

77. MOSANDL, A., PALM, U., GÜNTHER, C., KUSTERMANN, A. (1989) Z. Lebensm. Unters. Forsch. **188**: 148

78. SINGER, G., HEUSINGER, G., FRÖHLICH, O., SCHREIER, P., MOSANDL, A. (1986) J. Agric. Food Chem. **34**: 1029

79. SCHURIG, V., OSSIG, J., LINK, R. (1989) Angew. Chem. **101**: 197

80. KOSCIELSKI, T., SYBILSKA, D., JURCZAK, J. (1983) J. Chromatogr. **280**: 131

81. SCHURIG, V., NOWOTTNY, H.-P. (1988) J. Chromatogr. **441**: 155

82. KÖNIG, W. A., LUTZ, S., WENZ G., V. D. BEY, E. (1988) J. High Res. Chromatogr. **11**: 506

83. KÖNIG, W. A., LUTZ, S., COLBERG, C., SCHMIDT, N., WENZ, G., V. D. BEY, E., MOSANDL, A., GÜNTHER, C., KUSTERMANN, A. (1988) J. High Res. Chromatogr. **11**: 621

84. KÖNIG, W. A. (1989) Nachr. Chem. Tech. Lab. **37**: 471

85. MOSANDL, A., KUSTERMANN, A., PALM, U., DORAU, H.-P., KÖNIG, W. A. (1989) Z. Lebensm. Unters. Forsch. **188**: 517

86. MOSANDL, A., KUSTERMANN, A., HENER, U., HAGENAUER-HENER, U. (1989) Deutsche Lebensm. Rundschr. 205

87. TRESSL, R., HEIDLAS, J., ALBRECHT, W., ENGEL, K.-H. (1988) in "Bioflavour '87", (Schreier, P., Ed.), Walter de Gruyter, Berlin, New York, p. 221

88. ENGEL, K.-H., HEIDLAS, J., ALBRECHT, W., TRESSL, R. (1989) in "Flavor Chemistry Trends and Developments" (Teranishi, R., Buttery, R. G., Shahidi, F., Eds.), ACS Symposium Series, American Chemical Society, Washington, D. C., p. 8

89. FREYTAG, W., NEY, K. H. (1968) Europ. J. Biochem. **4**: 315

90. TRESSL, R., BAHRI, D., ENGEL, K.-H. (1982) J. Agric. Food Chem. **30**: 89

91. WURZENBERGER, M., GROSCH, W. (1983) Z. Lebensm. Unters. Forsch. **176**. 16

92. WURZENBERGER, M., GROSCH, W. (1984) Biochim. Biophys. Acta **795**: 163

93. ENGEL, K.-H., BUTTERY, R. G., MON, T. R., TERANISHI, R. (1989) J. Agric. Food Chem., submitted for publication

94. ENGEL, K.-H., FLATH, R. A., BUTTERY, R. G., TERANISHI, R. (1989) J. Agric. Food Chem., submitted for publication

95. MOSANDL, A., HEUSINGER, G., GESSNER, M. (1986) J. Agric. Food Chem. **34**: 119

96. FRÖHLICH, O., HUFFER, M., SCHREIER, P. (1989) Z. Naturforsch., **44c**: 555

97. MOSANDL, A., DEGER, W. (1987) Z. Lebensm. Unters. Forsch. **185**: 379

98. GERLACH, D., MISSEL, C., SCHREIER, P. (1988) Z. Lebensm. Unters. Forsch. **186**: 315

99. ENGEL, K.-H., BOHNEN, M., TRESSL, R. (1989) Z. Lebensm. Unters. Forsch. **188**: 144

100. TRESSL, R., ENGEL, K.-H. (1983) in "Instrumental Analysis of Foods" (Charalambous, G., Inglett, G., Eds.), Vol. 1, Academic Press, New York, p. 153

101. ENGEL, K.-H., TRESSL, R. (1983) J. Agric. Food Chem. **31**: 796

102. TAKEOKA, G., BUTTERY, R. G., FLATH, R. A., TERANISHI, R., WHEELER, E. L., WIECZOREK, R. L., Güntert, M. (1989) in "Flavor Chemistry Trends and Developments" (Teranishi, R., Buttery, R. G., Shahidi, F., Eds.), ASC Symposium Series, American Chemical Society, Washington, D. C., p. 223

103. ENGEL, K.-H., HEIDLAS, J., TRESSL, (1990) in "Food Flavors, Vol. 3: The Flavor of Fruits" (MacLeod, A. J., Morton, I. D., Eds.), Elsevier Publications, Amsterdam, p. 195

104. HEIDLAS, J., ENGEL, K.-H., TRESSL, R. (1988) Eur. J. Biochem. **172**: 633

105. TRESSL, R., ENGEL, K.-H., ALBRECHT, W. (1988) in "Adulteration of Fruit Juice Beverages" (Nagy, S., Attaway, J. A., Rhodes, M. E., Eds.), Marcel Dekker, New York, Basel, p. 67

106. FEUERBACH, M., FRÖHLICH, O., SCHREIER, P. (1988) J. Agric. Food Chem. **36**: 1236

107. GÜNTHER, C., MOSANDL, A. (1987) Z. Lebensm. Unters. Forsch. **185**: 1

108. MAARSE, H., VIISHER, C. A. (Eds.), (1984) "Volatile Compounds in Food, Qualitative Data", CIVO-TNO, Zeist

109. ENGEL, K.-H., FLATH, R. A., BUTTERY, R. G., MON, T. R., RAMMING, D. W., TERANISHI, R. (1988) J. Agric. Food Chem. **36**: 549

110. ENGEL, K.-H., RAMMING, D. W., FLATH, R. A., TERANISHI, R. (1988) J. Agric. Food Chem. **36**: 1003

II.5 High Molecular Weight Compounds

R. Hardt

1 Introduction

Gas chromatography (GC) is a well established method in the analysis of volatile low molecular weight compounds in food. Especially, capillary gas chromatography (HRGC) is the method of choice in many problems of food analysis, because of its extremely high separation efficiency. Compounds to be analyzed by GC have to be volatile without decomposition. Many food components, however, have high molecular weights or low thermal stability. This chapter outlines the possibilities for the analysis of such compounds by HRGC.

2 Theoretical Aspects

2.1 High Temperature Gas Chromatography

In general, the volatility of chemical compounds is decreased with an increase in molecular weights. Consequently, elevated temperatures have to be used in the GC of high molecular weight compounds. The limiting factors in high temperature gas chromatography (HTGC) are the maximum operating temperatures of the GC and the capillary column, as well as the thermal stability of the compounds to be analyzed. The development of metal-clad fused silica columns coated with cross-linked and chemically bonded stationary phases of high thermal stability allows operation at temperatures as high as 500 °C [1, 2], and 430 °C using columns with a high temperature polyimide coating. Recently, some manufacturers of gas chromatographic equipment offer instruments which are capable of column oven temperatures up to 520 °C.

Such elevated temperatures have been used in the gas chromatographic analysis of long chain hydrocarbons, which are of relatively high thermal stability, however, less stable compounds, such as many food ingredients, may thermally decompose at temperatures required for their elution.

2.1.1 Injection Techniques

The sampling of high molecular weight compounds requires careful attention in order to avoid quantitative sample losses. Low discrimination sample introduction is now generally achieved using techniques such as cold on-column and PTV injection systems. In contrast the split injection has a reputation of showing severe discrimination of high-boiling compounds. However, DAWES and CUMBERS [1] have shown, that split injection can be used in HTGC of high molecular weight compounds (n-alkanes up to C_{80}) without discrimination when compared to the cold on-column technique.

Significant losses of compounds above n-C_{60} have also been observed with programmed temperature splitless injection, when the PTV inlet is purged too soon after injection. HINSHAW and ETTRE [3] have demonstrated, that a split flow turn-on time of one minute leads to a significant discrimination of high-boiling substances above C_{62}. The pattern of discrimination shifts towards higher carbon numbers as the split turn-on time increases. The results for PT split injection and cold on-column injection were equivalent to ± 5 %.

PTV injectors contain a glass liner packed with some material (e.g. silanized quartz wool) to provide efficient transfer of liquid sample from the syringe to the injector. Recently a new injection system for HTGC, the Septum-equipped Programmable Injector (SPI) has been developed [4]. This injector contains an especially designed glass insert, which allows operation in a mode analogous to on-column injection. However, a special syringe needle is not required in PTV injection technique, and automatic injection is readily achieved with many types of autosamplers.

Some problems may arise in the analysis of high molecular weight substances with an on-column injector by non-volatile sample by-products. For these the use of an uncoated, deactivated pre-column, a so-called retention gap, is generally recommended. The pre-column prevents the capillary column to be contaminated by low volatile residues. Furthermore, the use of a retention gap reconcentrates the solute bands at the beginning of the coated column [5]. Band length must be as short as possible at the beginning of the chromatographic process to fully utilize the separation efficiency of the capillary column [6].

The risk of column contamination is diminished by using a PTV injector. Non-volatile residues are accumulated in the removable liner instead of in the capillary column itself.

2.1.2   Chromatographic Conditions

Not only may the chosen injection technique lead to quantitative sample losses, but also the chromatographic conditions. The exposure of thermally labile sample molecules to the elevated operating temperatures in HTGC raises the possibility of rearrangement or decomposition reactions during the chromatography. The probability of such thermal reactions decreases with lower temperatures and shorter exposure times of the sample to high temperatures.

Analysis time and elution temperature may be affected by varying the following parameters [3, 6]: column length and diameter, polarity and thickness of the stationary phase, carrier gas velocity, and the temperature program of the oven. All of these parameters have to be optimized to obtain the desired degree of resolution power in combination with an as short as possible analysis time. Relatively short columns with thin apolar films have been used successfully for a large number of HTGC analyses. HINSHAW and ETTRE [3] have investigated thermal effects in the column for a series of pure triglycerides by varying the temperature program rate and initial carrier gas velocity. The elution temperature of a compound is increased significantly by higher program rates and lower gas velocities. However, little effect on triglyceride recoveries was observed by changing the temperature programming rate. In contrast, changing the initial gas velocity from 30 to 125 cm/s had a significant effect on recoveries of the later eluting peaks, as shown in Figure II.5.–1. Greatly improved recoveries were obtained at the higher gas velocity. The described effect was independent of the injection technique chosen. Equal recoveries were observed for PT split, PT splitless, and cold on-column injection. Decomposition reactions while traversing the column may be the reason for these sample losses, although expected baseline fluctuations caused by elution of breakdown products were not observed.

2.2 Derivatization

Chemical derivatization is a very useful tool for the manipulation of samples when the boiling points of the components are too high for HTGC, or there are polar functional groups present that prevent chromatography due to strong intermolecular hydrogen bonding. This method is not only used to increase the volatility of sample components, but also to enhance their thermal stability. The sensitivity of a chromatographic method may be significantly improved by adding functional groups that allow the use of specific detectors such as electron capture. Furthermore, chemical derivatization may improve the separation of components that are not easily differentiated in the underivatized sample such as isomeric compounds. The improvement of peak symmetry by derivatization of active species such as hydrogen atoms leads to an increase in response. Derivatization reactions for GC mainly involve esterification, alkylation, or silylation [7]. Some examples for this method will be presented later in this chapter.

2.3 Chemical Decomposition

The direct analysis of extremely high molecular weight compounds such as proteins and polysaccharides by GC is impossible. These large molecules must be broken down to smaller products before GC can be attempted. In most cases polymeric substances are chemically broken down into their monomeric units. Enzymatic degradation techniques may also be used. The decomposition products can then either be directly analyzed by GC, or if necessary after a chemical derivatization step.

2.4 Pyrolysis

In pyrolysis, high molecular weight compounds to be analyzed are thermally fragmented in an inert atmosphere to form products of lower molecular weight and higher volatility. This method has been widely used in the analysis of synthetic polymers and biological materials. IRWIN [8] has given an excellent overview upon the different pyrolysis techniques and their applicability. Pyrolysis methods can be classified to two major groups depending upon the heating mechanism, either the continuous mode or the pulse mode.

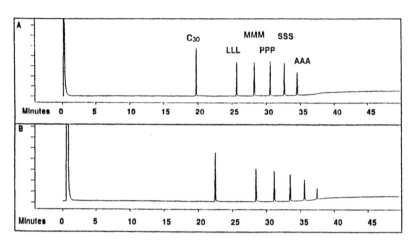

Fig. II.5.–1 Synthetic triglyceride standard at low and high linear velocities. **A** 125 cm/s **B** 30 cm/s [3].

Conditions: 7 m × 0.25 mm Al-clad fused silica WCOT, coated with high temperature methylsilicone (0.15 μm). *Temperature:* 50 °C initial, 10 °C/min to 420 °C. *Carrier gas:* He. *Injection:* PTV, 50 °C initial, 440 °C final, split flow: 50 ml/min. *Detector:* FID, 450 °C.

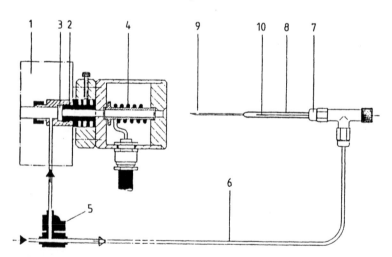

Fig. II.5.–2 Curie-point pyrolysis unit for mounting in front of the GC-injector [16].

1 GC, **2** injector, **3** septum, **4** high frequency induction coil, **5** solenoid valve, **6** PTFE carrier gas tube, **7** T-connection, **8** glass pyrolysis chamber, **9** stainless steel injection needle, **10** ferromagnetic conductor.

In continuous mode systems such as microreactors or furnace pyrolyzers the sample container is heated by an external source normally a metal tube which is energized by an electrical resistive heating device. A common method is to introduce the sample, usually up to milligram amounts in a "boat", into the tube free from wall contact.

The two most common techniques in the pulse mode are resistive heating of a platinum filament (heated-filament pyrolysis), or inductive heating of a ferromagnetic conductor (Curie-point pyrolysis). In both methods the sample, usually in the microgram or sub-microgram range, is coated directly onto the wire, which is heated to the equilibrium temperature in just some milliseconds.

In pyrolysis gas chromatography (PyGC), the sample to be analyzed is thermally fragmented into a stream of an inert carrier gas, and the reaction products are passed directly into a gas chromatographic column where they are separated. Under the following conditions, the structure of the fragments obtained correlates closely with the structure of the starting material [9, 10]:

1. The fragmentation of the sample must be carried out in such a high dilution and the fragments so quickly stabilized that no undesirable secondary reactions occur.

2. The sample must be heated to a constant pyrolysis temperature as quickly and as reproducibly as possible.

3. The pyrolysis temperature must not be too high, since non-characteristic fragments can otherwise be preferentially formed.

The above conditions are satisfied by Curie-point pyrolysis GC in a nearly ideal manner [9–15]. In this method a ferromagnetic conductor in contact with the sample is heated in 20 to 30 milliseconds inside a low dead volume glass pyrolysis chamber which is located along the axis inside a high frequency coil. The power consumption of the conductor depends on the magnetic field inside the coil, the radius, specific resistance, and magnetic permeability of the conductor, as well as the frequency applied. Owing to the drastic change in the magnetic permeability of the ferromagnetic conductor at the Curie-point, the energy input drops at this temperature. If the energy loss of the conductor above the Curie-point is equal to or larger than the energy uptake from the field, the conductor will warm up only to that temperature. By using conductors of different materials the Curie-point can be varied from 300 to 900 °C.

Figure II.5.–2 shows schematically a Curie-point pyrolysis unit (Model 310, Fischer Labor-und Verfahrenstechnik, Meckenheim b. Bonn, F.R.G.) attached to a GC. The ferromagnetic conductor coated with the sample is placed into a glass pyrolysis chamber equipped with a stainless steel injection needle. For pyrolysis, the needle is inserted through the septum of the GC injector in such a way that the conductor is centered axially in the high frequency induction coil which is fitted to the injector. By switching a solenoid valve carrier gas can be passed through a PTFE tube connected to the pyrolysis chamber. Pyrolysis time is variable in a range from 0.2 to 10 seconds.

Pyrolysis gas chromatography may be used as a fingerprint technique for the material studied, or as a method in the structural elucidation of organic compounds, as well as in the qualitative and quantitative analysis of mixtures. The pyrolysis gas chromatogram of a defined mixture can be obtained by adding together the signals of pyrograms of the individual components. The coupling of capillary gas chromatography with mass spectrometry (MS) or Fourier transform infrared spectroscopy (FTIR) is an attractive analytical tool for the separation of pyrolysis products and the subsequent determination of their structures.

3 Applications

3.1 Lipids

Recent reviews on HRGC applied to lipid analysis have been published by TRAITLER [17] and MARES̆ [18]. Therefore, only some important applications will be discussed in this chapter.

3.1.1 Triglycerides

Triglycerides containing higher molecular weight fatty acid groups such as stearic or oleic acid belong to the least volatile compounds in food which have been analyzed directly by gas chromatography. In contrast to the analysis of fatty acid methyl esters, GC of triglycerides not only enables the analyst to obtain information upon nature and amount of the fatty acids present in a fat or oil but also on their distribution in the component triglyceride.

Some years ago, only apolar stationary phases could be used in the GC of triglycerides because of the high temperatures (over 300 °C) needed for their elution. On apolar columns, the GC separation of triglycerides is based on the difference in carbon number. During the past few years thermally stable stationary phases (50 % phenyl–/50 % methyl-silicone) have been developed, so that resolution is now achieved not only by carbon number, but also by a further refined separation which is related to the degree of unsaturation. Polarity increases relative to the total number of double bonds present in a triglyceride molecule. As a result, those triglyceride molecules with a high proportion of double bonds have the longest retention times [19, 20]. Using capillary columns with such stationary phases, the triglycerides of a fat or oil can be separated in less than 30 minutes. For example, Figure II.5.–3 shows the analysis of the triglycerides of a butter fat, and Figure II.5.–4 the chromatogram of a peanut oil, respectively. As is clearly demonstrated, triglycerides of the same carbon number are eluted within a group, but with a further separation according to the number of double bonds.

Mass spectrometry is a very efficient tool in the structure determination of triglycerides. This method allows the identification of the fatty acids present in a triglyceride as well as

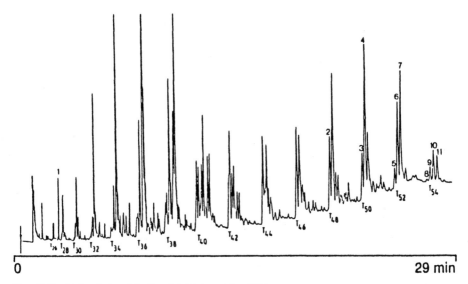

Fig. II.5.–3 Analysis of butter fat triglycerides [19].

> *Conditions:* 25 m × 0.25 mm armoured fused silica WCOT, coated with TAP
> (0.10 μm), Chrompack. *Temperature:* 280 °C (1 min) initial, 3 °C/min to
> 335 °C. *Carrier gas:* H$_2$, 100 kPa. *Injector:* on-column. *Sample:* 0.2 μl of
> 0.05 % butter fat in hexan. *Detector:* FID.
>
> *Peak identification:* **1** cholesterol, **2** PPP, **3** PPS, **4** PPO, **5** PSS, **6** PSO,
> **7** POO, **8** SSS, **9** SSO, **10** SOO, **11** OOO. **P** = palmitic acid, **S** = stearic
> acid, **O** = oleic acid

the elucidation of the position of the acid groups [21–27]. Here one has to consider that
co-eluting trigylcerides may lead to misinterpretations. It seems to be impossible to totally
separate all the triglycerides, that can be present in a fat or oil, because of the multitude
of possible triglyceride molecular species. If only the 10 most common fatty acid groups
are considered, a complete analysis would require the separation of up to 550 different
triglycerides.

GC of triglycerides is suitable as a method in the investigation of the purity of fats and oils,
and in the qualitative and quantitative analysis of fat mixtures. As an example, cocoa butter can be distinguished from cocoa butter substitutes, even when the substitutes are
composed of mixtures having nearly the same composition of fatty acids [28, 29]. Typical
gas chromatograms of cocoa butter and some different substitutes using an apolar
column are shown in Figure II.5.–5. The better separation of cocoa butter triglycerides of
the same carbon number on a more polar stationary phase is demonstrated in Figure
II.5.–6.

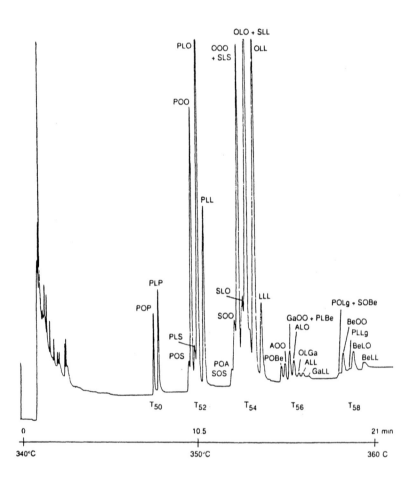

Fig. II.5.–4 Analysis of peanut oil triglycerides [20].

Conditions: 25 m × 0.25 mm WCOT, coated with RSL-300 (0.1 μm), Alltech. *Temperature:* 340 °C initial, 1 °C/min to 360 °C. *Carrier gas:* H_2, 0.8 bar. *Injector:* H.O.T. cold on-column. *Sample:* 0.2 μl of 0.05 % peanut oil in iso-octan. *Detector:* FID.

Peak identification: **P** = palmitic acid, **S** = stearic acid, **A** = arachidic acid, **Be** = behenic acid, **Lg** = lignoceric acid, **O** = oleic acid, **L** = linoleic acid, **Ga** = gadoleic acid.

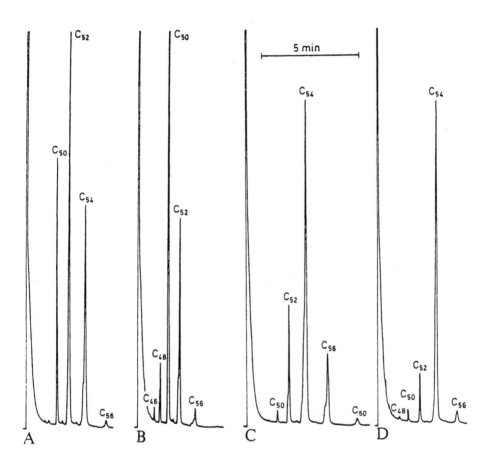

Fig. II.5.–5 Triglycerides of cocoa butter **A** and cocoa butter substitutes. Calvetta **B**, Sal
fat **C**, Illexao **D**. From [30].

Conditions: 8 m × 0.3 mm glass WCOT, coated with OV-101. *Temperature:*
330 °C isothermal. *Carrier gas:* H$_2$, 0.5 bar. *Injection:* split 1:10.

Thermal degradation of fats has practical importance because it takes place in cooking
and food processing. The pyrolytic behaviour of triglycerides was investigated by several
authors [32–39]. The intention of these works was primarily to determine, which products
are formed by heating triglycerides in the absence of air, or to get information on the
degradation mechanisms. In some cases the triglycerides have been heated for several
hours, and the pyrolysis products formed were analyzed by GC using packed columns. In

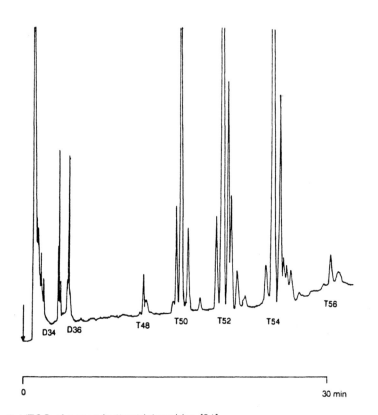

Fig. II.5.–6 HTGC of cocoa butter triglycerides [31].

     Conditions: 25 m × 0.25 mm fused silica WCOT, coated with TAP (0.1 μm), Chrompack. *Temperature*: 340 °C initial, 1 °C/min to 365 °C. *Carrier gas*: H_2 100 kPa. *Injector*: cold on-column. *Detector*: FID, 380 °C. *Sample*: 0.1 μl of 0.05 % cocoa butter in pentane.

Curie-point pyrolysis HRGC of tristearin nearly 90 pyrolysis products could be identified [40]. Figure II.5.–7 shows the pyrograms of tristearin obtained at different pyrolysis temperatures. The most prominent pyrolysis products identified by mass spectrometry are homologous series of 1-alkenes and n-alkanes with 1-heptadecene (a) and n-heptadecane (b) as the homologs with the highest chain length, and stearic acid (c). The fragment with the highest degree of structural information is 2-propenyl stearate, a compound

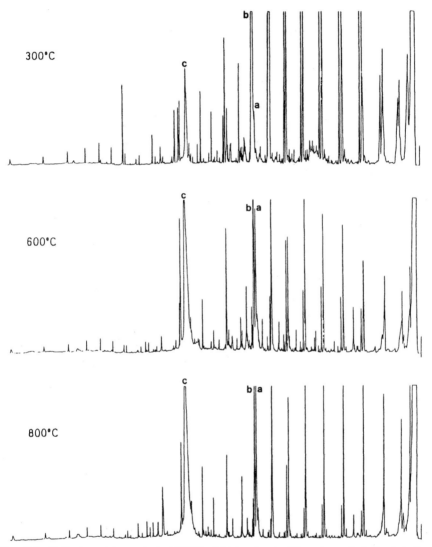

Fig. II.5.–7 Curie-point pyrolysis HRGC of tristearin at different pyrolysis temperatures
[40].

Conditions: Fischer Curie-point pyrolyzer Mod. 310. Pyrolysis time: 10 s.
Sample: 10 μg of tristearin. 25 m × 0.25 mm glass WCOT, coated with SE-
30 (0.19 μm), Chrompack. Temperature: 40 °C (5 min) initial, 3 °C/min to
270 °C. Carrier gas: He, 2 ml/min. Injector: split-/splitless-injector, 300 °C,
1 min splitless. Detector: FID, 300 °C.

containing the parent acid as well as the three carbon atoms of the glycerol part of the tri-gylceride molecule. By means of these fragments, the identification of a mixed triglyceride should also be possible. Whether or not the position of the fatty acids in the glyceride molecule can be determined by PyGC, as it is possible by mass spectrometry, is still a question. Aldehydes, alkylfurans, alkylbenzenes, and some different bi- and tricyclic aromatic hydrocarbons belong to the pyrolysis products detected in smaller amounts. All of these fragments are already formed at a pyrolysis temperature as low as 300 °C. This temperature is exceeded in HTGC of triglycerides, so that thermal decomposition reactions may also occur using this method.

3.1.2　Sterols

In many cases HRGC of sterols is an excellent method in the characterization of fats and fat mixtures. Adulterations of animal fats with vegetable fats, and vice versa are easily detectable. Mixtures of vegetable fats are also testable [41]. For analysis, the sterols are being extracted from a fat after an initial saponification. Thin layer chromatography, liquid chromatography, or lipophilic gel chromatography have been used as subsequent clean-up procedures [42, 43]. The extract containing the sterols can directly be analyzed by GC, or after derivatization of the OH-groups. Figure II.5.–8 shows the chromatogram of a mix-ture of free plant sterols, and Figure II.5.–9 that of TMS derivatives of sterols in tuna olive oil.

3.2　Proteins, Peptides, and Amino Acids

Direct gas chromatography of proteins is impossible, because of their extremly high molecular weights, extensive hydrogen bonding, and ionic character. Proteins must be reduced to their component amino acids by acidic or basic hydrolysis, before GC can be attempted. The carboxyl as well as the amino groups of the amino acids must be deri-vatized to achieve sufficient volatility and stability for GC analysis. Modification of the car-boxyl groups to their propyl or butyl esters, and derivatization of the amino groups using trifluoroacetic (TFA), heptafluorobutyric (HFB), or acetic anhydride, is a common method [41, 42].

At the present time amino acid analyzers and high-performance liquid chromatography (HPLC) methods are predominantly used in the determination of the amino acid compo-sition of food, while GC is only of secondary importance. However, HRGC is excellently suitable in the separation of the optical isomers of amino acids. Figure II.5.–10 shows the GC separation of the enantiomers of 19 amino acids using a chiral phase capillary column.

The thermal fragmentation of amino acids was investigated in detail by several authors. Already in 1965 SIMON and GIACOBBO showed in one of their fundamental papers on Curie-point pyrolysis GC, that the naturally occurring amino acids can be identified on the

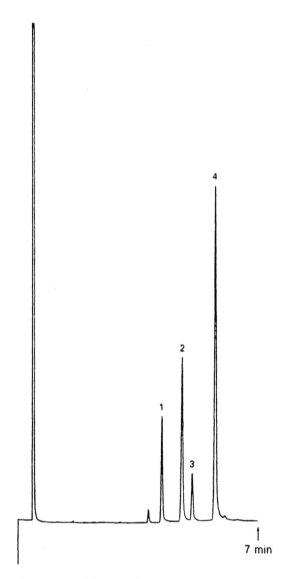

7 min

Fig. II.5.–8 Analysis of a mixture of free plant sterols [44].

Conditions: 15 m × 0.25 mm fused silica WCOT, coated with DB-1701 (0.15 μm), J&W. *Temperature*: 260 °C isothermal. *Carrier gas*: H₂, 55 cm/s. *Injector*: split. *Detector*: FID.

Peak identification: **1** brassicasterol, **2** campesterol, **3** stigmasterol, **4** β-sitosterol.

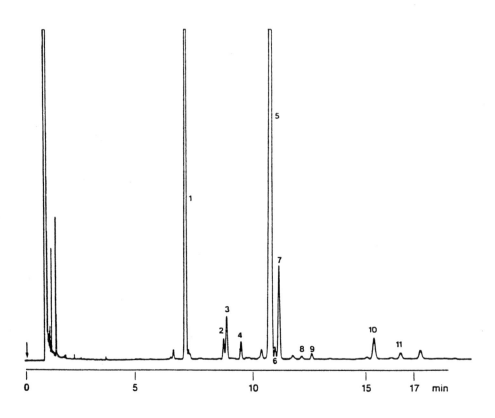

Fig. II.5.–9 TMS-Derivatives of sterols in tuna olive oil [45].

    Conditions: 15 m × 0.22 mm fused silica WCOT, coated with CP-Sil 8 CB (0.12 μm), Chrompack. *Temperature*: 240 °C isothermal. *Carrier gas*: H_2, 130 kPa. *Injection*: split 1:60. *Detector*: FID.

    Peak identification: **1** cholesterol, **2** 24-methylenecholesterol, **3** campesterol, **4** stigmasterol, **5** β-sitosterol, **6** sitosterol, **7** δ5-avenasterol, **8** δ7-stigmasterol, **9** δ7-avenasterol, **10** erythradiol, **11** uvaol.

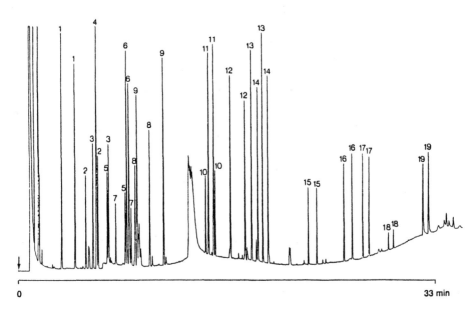

Fig. II.5.–10 HRGC analysis of amino acids (TFA/isopropyl esters) [46].

Conditions: 25 m × 0.22 mm fused silica WCOT, coated with Chirasil-L-VAL (0.12 μm), Chrompack. Temperature: 75 ° (3 min) initial, 3 °C/min to 195 °C. Carrier gas: H_2, 90 kPa, 48 cm/s. Injector: split, 30 ml/min, 220 °C. Detector: FID, 220 °C. Sample: 0.05 μl, 1–2 ng/component.

Peak identification: **1** DL-ALA, **2** DL-VAL, **3** DL-THR, **4** GLY, **5** DL-α-ILE, **6** DL-PRO, **7** DL-ILE, **8** DL-SER, **9** DL-LEU, **10** DL-CYC, **11** DL-ASP, **12** DL-MET, **13** DL-PHE, **14** DL-GLU, **15** DL-TYR, **16** DL-ORN, **17** DL-LYS, **18** DL-ARG, **19** DL-TRP. D-enantiomer elutes first.

basis of their pyrograms. They used 30 to 50 m long GOLAY columns to separate the pyrolysis products. The fragments observed in the pyrolysis of amino acid mixtures were also formed in the fragmentation of the corresponding dipeptides, but the relative intensities of the signals were affected by the peptide linkage [9]. The coupling of PyGC with mass spectrometry enabled the identification of the pyrolysis products [10, 12], and information about the fragmentation mechanisms could be obtained by pyrolysis radio gas chromatography of [14]C- and [3]-H-labelled amino acids [10].

Important pyrolysis products of amino acids with alkyl or aryl residues are the corresponding nitriles, formed by decarboxylation of the amino acids and subsequent dehydration of the resulting amines. Alkanes and alkenes, representing the alkyl residues, as well as aldehydes and ketones have also been found in the pyrolyzates of aliphatic amino acids

[12, 47–49]. On pyrolysis of aromatic amino acids, the fragmentation of the aromatic nucleus is negligible [12, 50]. Nitrogen-containing heterocyclic compounds, such as pyridines, piperidines, and pyrroles belong to the pyrolysis products of lysine [51].

3.3 Carbohydrates and Related Compounds

3.3.1 Mono-, Oligo-, and Polysaccharides

Carbohydrates belong to the main ingredients in food of plant origin. Different methods are used in their analysis, especially enzymatic analysis, HPLC, and GC particularly in the determination of mono- and oligosaccharides. Enzymatic analysis is well established and specific, but the specificity can be a drawback when several sugars are to be assayed. In such cases this method becomes time consuming and expensive. GC and HPLC are more common in the qualitative and quantitative analysis of complex carbohydrate mixtures. Since carbohydrates have a low volatility and lack thermal stability, they have to be derivatized prior to GC analysis, however chemical derivatization of the hydroxyl groups is not required in HPLC. HRGC is much more sensitive than HPLC with refractive index (RI) detection, as well as the separation power of HRGC is higher when compared with HPLC [52].

Before analysis, polysaccharides, which are often used as thickening agents in food, must be decomposed to their monomeric units enzymatically or by acid hydrolysis. Pyrolytic methods have also been used with success for the characterization of polysaccharides.

3.3.1.1 HRGC of Volatile Derivatives

Different techniques have been applied to obtain derivatives of carbohydrates having volatility and stability high enough for GC analysis. For example, sugars have been converted to their methyl ethers, acetates, trimethylsilyl (TMS) ethers, or trifluoroacetates. Anomerization of reducing sugars may lead to multiple peaks, and can complicate the chromatograms. Co-eluting peaks make quantitative analysis difficult as well. A simplification of the chromatograms can be achieved by sodium borohydride reduction of the carbonyl group of reducing sugars to form the corresponding alditols, which can be analyzed as their TMS derivatives or acetates, although information may be lost because some sugars yield the same alditol [41, 42, 53].

Low and Sporns [53] described a method to separate and quantitate most of the di- and trisaccharides composed of glucose and fructose moieties found in honey. The oligosaccharides in honey, representing approximately 3 % of the total sugars, can consist of more than twenty different di- and trisaccharides. At least twelve closely related di- and seven trisaccharides were successfully identified and quantitated employing a combination of reduction and trimethylsilylation. Figure II.5.–11 shows the capillary gas chromatogram of the disaccharides, and Figure II.5.–12 that of the trisaccharides in an alsike honey

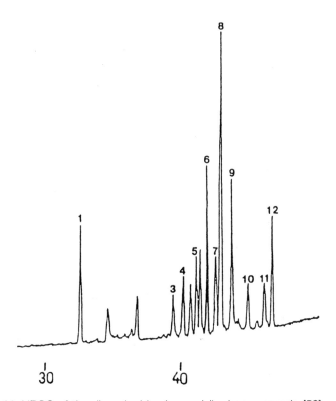

Fig. II.5.–11 HRGC of the disaccharides in an alsike honey sample [53].

Conditions: 30 m × 0.25 mm fused silica WCOT, coated with DB-5, J&W.
Temperature: 210 °C (12 min) initial, 2 °C/min to 290 °C. *Carrier gas*: He,
27 cm/s. *Injection*: split 1:30, 300 °C. *Detector*: FID, 300 °C.

Peak identification: **1** sucrose, **3** neotrehalose, **4** cellobiose, **5** laminari-
biose, **6** turanose and nigerose, **7** maltulose and turanose, **8** maltose and
maltulose, **9** kojibiose, **10** gentiobiose, **11** palatinose, **12** isomaltose and
palatinose.

sample. Most of the disaccharides could be indentified and quantitated directly. An
exception was the disaccharide maltulose in which both reduction peaks were found to
overlap, with the glucitol peak identical to the maltose reduction peak, and the mannitol
peak overlapping with the glucitol peak of turanose. The separation of these overlapping
peaks was made possible using isothermal conditions at 250 °C. Therefore two injections
were necessary for the quantitation of all disaccharides present in this honey.

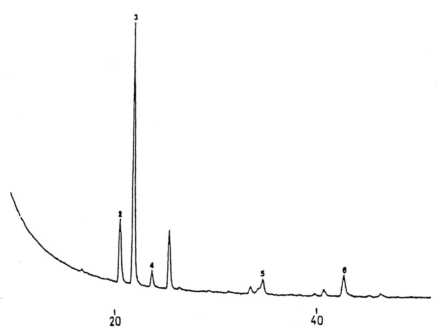

Fig. II.5.–12 HRGC of the trisaccharides in an alsike honey sample [53].

Temperature: 290 °C isothermal. Other conditions as in Figure II.5.–11.

Peak identification: **2** isopanose, **3** erlose, **4** theanderose, **5** maltotriose, **6** panose.

Procedures for the simultaneous determination of a wide variety of monosaccharides ranging from five-carbon atoms through nine-carbon atoms and consisting of neutral, alcohol, and amino sugars have been described by GUERRANT and MOSS [54]. Neutral and amino sugars were analyzed as their aldonitrile acetates, while alcohol sugars were simultaneously derivatized as alditol or cyclitol acetates. Aldonitrile acetate derivatives have been used instead of alditol acetate derivatives because of easier preparation, greater stability, and good chromatographic separation. In addition, neutral and amino sugars were derivatized as O-methyloxime acetates, with alcohol sugars as alditol or cyclitol acetates.

SCHERZ and MERGENTHALER [55] have reviewed the analytical methodology for the determination of polysaccharides, used as thickening agents in foods. After isolation and purification, the polysaccharides are chemically degraded to their monomeric units by acid hydrolysis. The monosaccharides can then be identified and quantitated by gas chroma-

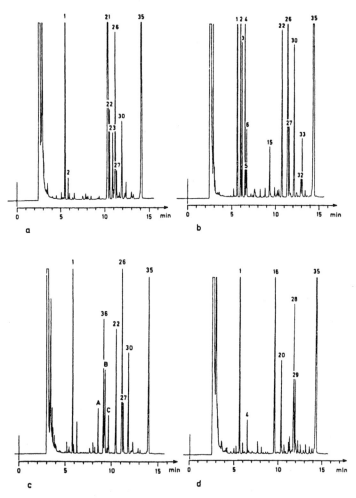

Fig. II.5.–13 HRGC of the monomeric units of some polysaccharides, used as thickening agents in food [57].

a locust bean gum, **b** gum arabic, **c** agar, **d** pectin.

Conditions: 35 m × 0.3 mm glass WCOT, coated with SE-30. *Temperature*: 120 °C initial, 4 °C/min to 190 °C. *Carrier gas*: N$_2$, 1.5 ml/min. *Injection*: split 1:20, 250 °C. *Detector*: FID, 250 °C.

Peak identification: **1** isoerythrit (internal standard), **2 3 6** arabinose, **4 5** rhamnose, **15 32 33** glucuronic acid, **16 20 28 29** galacturonic acid, **21 23** mannose, **22 26 27 30** galactose, **35** sorbitol (internal standard), **A B C** 6-methylgalactose.

tography preferably as their aldonitrile acetates. Using this derivatization method, each monosaccharide leads to a single peak in the chromatogram.

Polysaccharide mixtures having identical sugar units, but differing in their linkages can be identified by means of methylation analysis [55, 56]. Such mixtures are methylated prior to hydrolysis, and the resulting methyl sugars are separated and identified by GC as their aldonitrile acetate derivatives.

PREUSS and THIER [57, 58] have presented an analytical method for fast and simple isolation of natural thickeners and gums from a great variety of foods and the subsequent quantitative determination by HRGC. Methanolysis was used instead of hydrolysis to decompose the polysaccharides. The methylglycosides formed were analyzed as their trimethylsilyl ethers. Figure II.5.–13 shows the capillary gas chromatograms of the monomeric units of four different food thickeners analyzed by this method.

3.3.1.2 Pyrolysis

Heating of sugars leads to the formation of brown polymeric substances and a large number of volatile degradation products having a great aroma potential. 1,4:3,6-Dianhydro-D-glucopyranose, furans, aldehydes, ketones, diketones, and aromatic hydrocarbons belong to the main pyrolysis products of D-glucose [59]. HEYNS and KLIER [60] have shown, that on pyrolysis at 300 to 500 °C for a short period of time, different mono-, oligo-, and polysaccharides give the same volatile degradation products. They suggested, that all of these compounds form similar polymeric intermediates by degradation, dehydration, and condensation reactions, which undergo a secondary thermal fragmentation.

JOHNSON et al. [61] found, that the volatile products from sucrose pyrolysis and from an aqueous acid-stannous chloride degradation of glucose are similar in composition, which suggest some similarity in reaction mechanisms. PREY et al. [62] investigated systematically the amount of reaction water and of volatile organic fragments formed from D-glucose at different pyrolysis conditions. Results supported the assumption that the polymeric product being formed must consist of furan compounds.

In Curie-point pyrolysis HRGC-MS of different mono-, oligo-, and polysaccharides more than 125 fragmentation products could be identified [63]. Furan derivatives belong to the most concentrated decomposition products at a pyrolysis temperature of 700 °C. VAN DER KAADEN et al. [64] investigated the influence of inorganic additives to the pyrolysis mixture by Curie-point pyrolysis MS and Curie-point pyrolysis HRGC-MS of amylose. Carbonyl compounds, acids, and lactones were released from alkaline and neutral mixtures. Furans and anhydrohexoses are especially formed under neutral and acidic conditions. Pyranones have been found as being specific for phosphate matrices. Unsaturated hydrocarbons and aromatic substances arise from strongly alkaline or dehydrating matrices.

Identification of a single food thickener is possible by pyrolysis HRGC-MS [65]. Selected ion monitoring of peaks at m/z 60 (acids), 88 (ethyl esters), 81 (furans), 95 (pyranes), 108, and 126 were used to reveal differences between the tested polysaccharides. Many of the identified pyrolysis products were present in the pyrograms of all investigated food thickeners, but the differences in quantities of various compounds could be used in the identification of the polysaccharides. However, accurate analysis of a mixture of food thickeners was not achieved by this method.

HELLEUR [66] has shown that under optimum pyrolysis conditions polysaccharides will preferentially depolymerize by a transglycosidation reaction forming stable anhydro sugar products. The stereo-configuration of the initial saccharide units were thus retained. The anhydro sugars formed by pyrolysis were resolved on a capillary column and identified by mass spectrometry. It was found that uronic acid-containing polysaccharides required the carboxylate functional group to be protonated in order to obtain structurally-significant pyrolysis products.

3.3.2 Caramel Colours

Caramel colours are brown food colourings, produced by heating sugars in the presence of browning accelerators. On the basis of the accelerators used, the following classes of caramel colours are recognized [67]:

Class I   CP: Caramel Colour (Plain). Spirit Caramel, Non-ammonia Caramel

Class II   CCS: Caramel Colour (Caustic Sulphite Process)

Class III   AC: Ammonia Caramel Colour, Beer Caramel

Class IV   SAC: Sulphite Ammonia Caramel, Soft Drink Caramel, Acid Proof Caramel

At the present time gel filtration is used most often for the detection of caramel colours in foodstuffs. This procedure, however, is not specific and therefore cannot be employed on food containing brown compounds with high molecular weight naturally.

Curie-point pyrolysis HRGC-MS enables the differentiation between the four classes of caramel colours on the basis of the most concentrated of more than 100 identified pyrolysis products [68, 69]. Figure II.5.–14 shows the pyrograms of caramel colours of class I, III, and IV. The pyrogram of a caustic sulphite caramel colour and the mass fragmentogram of m/z 64 (SO_2, M^+) are presented in Figure II.5.–15. Sulphur dioxide is the main pyrolysis product of caramel colours of class II and IV. The pyrolyzates of caramel colours of class I, II, and IV contain primarily furans and furanones, while nitrogen-containing heterocyclic compounds, especially pyrazines, dominate the pyrograms of ammonia caramel colours. These compounds can also be found in lower amounts in the pyrolyzates of sulphite ammonia caramels. Their main precursors are polyhydroxyalkylpyrazines, which are ingredients of ammonia and sulphite ammonia caramels [70].

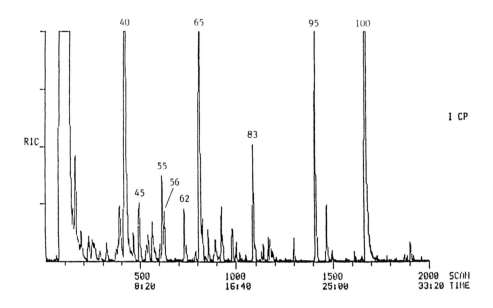

Fig. II.5.–14 Pyrolysis-gas chromatograms of caramel colours, class I, III, and IV (top to bottom) [68].

Conditions: Fischer Curie-point pyrolyzer Mod. 310. *Pyrolysis temperature*: 600 °C. *Pyrolysis time*: 10 s. *Sample*: 100 µg of caramel colours. 30 m × 0.25 mm fused silica WCOT, coated with DB-210 (0.25 µm), J&W. *Temperature*: 40 °C (5 min) initial, 3 °C/min to 210 °C. *Carrier gas*: He, 35 cm/s. *Injector*: split-/splitless-injector, 250 °C, 1 min splitless. *Detector*: MS. *Ionization*: EI, 70 eV.

Peak identification: **19** pyrazine, **20** pyridine, **27** methylpyrazine, **32** 2,5- and 2,6-dimethylpyrazine, **40** furan-2-carboxaldehyde, **45** 2-cyclopentene-1-one, **55** 2-acetylfuran, **56** benzaldehyde, **62** 5-methylene-2(5H)-furanone, **63** 5- and 6-methylpyrazine-2-carboxaldehyde, **65** 5-methylfuran-2-carboxaldehyde, **83** 2(5H)-furanone, **92** 2-pyridinecarbonitrile, **95** furan-2,5-dicarboxaldehyde, **100** 5-hydroxymethylfuran-2-carboxaldehyde.

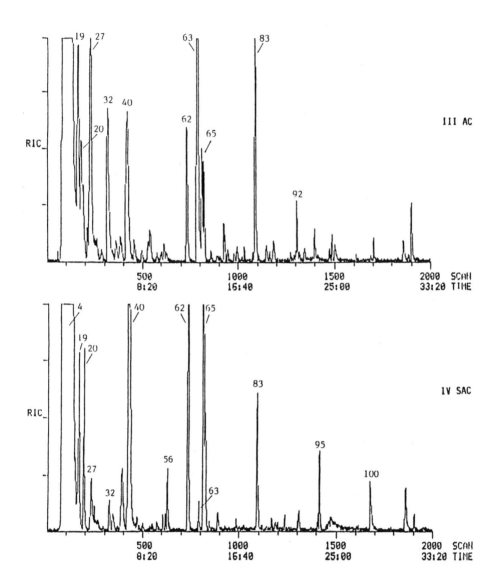

Fig. II.5.–14 Pyrolysis-gas chromatograms of caramel colours, class I, III, and IV (top to bottom) *(continued)*

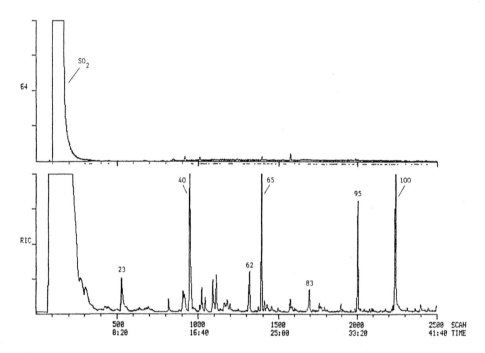

Fig. II.5.–15 Pyrolysis-gas chromatogram and mass fragmentogram of m/z 64 (SO$_2$, M$^+$) of a caramel colour, class II [68].

Conditions: 30 m × 0.32 mm fused silica WCOT, coated with DB-210 (0.50 µm), J&W. Other conditions as in Figure II.5.–14.

Peak identification: **23** 1-hydroxy-2-propanone. Other peaks as in Figure II.5.–14.

The detection of caramel colours in liquid foodstuffs such as vinegar, brandy, and soft drinks is also possible by Curie-point pyrolysis HRGC-MS after isolation of the polymeric ingredients of the colourings by ultrafiltration [71]. Curie-point pyrolysis of non-ammonia caramels and of caramel syrups, produced by heating sugars in the absence of browning accelerators, leads to the same degradation products. Such caramel syrups are used in foods because of their typical aroma. Light brown caramel syrups can be differentiated from caramel colours of class I CP on the basis of quantitative differences of some pyrolysis products, while extremely dark brown products, having an unpleasent aroma, yield the same fragments in nearly the same amounts as non-ammonia caramel colours [72].

4 Conclusions

Techniques as HTGC, chemical derivatization, chemical decomposition, or pyrolysis enable analyses of high molecular weight or thermally labile compounds by high resolution gas chromatography. When compared with HPLC, a method often used in the analysis of such compounds, HRGC offers more separation power and, using short capillary columns, a faster analysis time. The specificity and sensitivity of the detection techniques available are important factors in the choice of a chromatographic method. In this respect, GC has a clear advantage over HPLC. On the other hand, HPLC can be faster and more accurate, when hydrolysis or derivatization procedures are not needed. SANDRA and DAVID [73] have clearly illustrated the possibilities of different chromatographic techniques for a selected application.

Advantages of pyrolysis when compared to chemical degradation techniques are speed and simplicity of the method, and the low amount of sample needed, usually in the microgram range. However, chemical degradation can be the better method, when decomposition to the monomeric units is possible, and quantitative results are needed. MCKNIGHT HALKET and SCHULTEN [74] have described the analytical application of direct pyrolysis field ionization-mass spectrometry and Curie-point pyrolysis GC-MS to various whole foodstuffs. This methodology provides precise information from the sample in an extremely short period of time without any need of sample preparation, and offers great potential for the rapid characterization of foodstuffs and raw materials.

References

1. DAWES, P., CUMBERS, M. (1988) Proc. 9th Int. Symp. on Capillary Chrom. (Sandra, P., Ed.), Huethig, Heidelberg, p. 145

2. DESTY, D. H. (1987) J. Chromatogr. Sci. **25**: 552−563

3. HINSHAW, J. V., ETTRE, L. S. (1988) in c. f. [1], (Sandra, P., Ed.), p. 606

4. BERG J. R. (1988) in c. f. [1], (Sandra, P., Ed.), p. 695

5. GROB K., (1986) "Classical Split and Splitless Injection in Capillary Gas Chromatography. With Some Remarks on PTV Injection", Huethig Verlag, Heidelberg, Basel, New York, p. 166

6. JENNINGS, W. (1987) "Analytical Gas Chromatography", Academic Press, Orlando, San Diego, New York, Austin, Boston, London, Sidney, Tokyo, Toronto

7. Alltech − Applied Science, Deerfield, IL, USA, Bulletin No. 126 "GC Derivatization Guide"

8. IRWIN, W. J., (1982) "Applied Pyrolysis: A Comprehensive Guide", Marcel Dekker, New York

9. SIMON, W., GIACOBBO, H. (1965) Chem. Ing. Tech. **37**: 709−714

10. SIMON, W., KRIEMLER, P., VÖLLMIN, J. A., STEINER, H. (1967) J. Gas Chromatogr. **5**: 53−57

11. GIACOBBO, H., SIMON, W. (1964) Pharm. Acta Helv. **39**: 162−167

12. VÖLLMIN, J., KRIEMLER, P., OMURA, I., SEIBL, J., SIMON, W. (1966) Microchem. J. **11**: 73−86

13. BÜHLER, Ch., SIMON, W. (1970) J. Chromatogr. Sci. **8**: 323–329

14. WALKER, J. Q. (1972) Chromatographia **5**: 547–552

15. OERTLI, Ch., BÜHLER, Ch., SIMON, W. (1973) Chromatographia **6**: 499–502

16. Fischer Labor- und Verfahrenstechnik, Meckenheim b. Bonn, F.R.G. (1987) "Curie Point Pyrolysis and Automatic Sampler for GC-, IR- or MS-Application"

17. TRAITLER, H. (1987) Prog. Lipid Res. **26**: 257–280

18. MAREŠ, P. (1988) Prog. Lipid Res. **27**: 107–133

19. Chrompack International B. V., Middelburg, The Netherlands (1987) "Chrompack Packard Trigylceride Analyzer"

20. Alltech Europe, Eke, Belgium, Technical Information Bulletin No. 701 "The Analysis of Triglycerides on RSL-300"

21. RYHAGE, R., STENHAGEN, E. (1960) J. Lipid Res. **1**: 361–390

22. BARBER, M., MERREN, T. O., KELLY, W. (1964) Tetrahedron Lett. **18**: 1063–1067

23. SUN, K. K., HOLMAN, R. T. (1965) J. Am. Oil Chem. Soc. **45**: 810–817

24. LAUER, W. M., AASEN, A. J., GRAFF, G., HOLMAN, R. T. (1970) Lipids **5**: 861–868

25. AASEN, A. J., LAUER, W. M., HOLMAN, R. T. (1970) Lipids **5**: 869–877

26. ZEMAN, A., SCHARMANN, H. (1973) Fette, Seifen, Anstrichm. **75**: 170–180

27. NATALE, N. (1977) Lipids **12**: 847–856

28. FINCKE, A. (1976) Dtsch. Lebensm. Rundsch. **72**: 6–12

29. SCHULTE, E., FINCKE, A. (1978) Lebensmittelchem. Gerichtl. Chem. **32**: 15

30. SCHULTE, E. (1983) "Praxis der Kapillar-Gaschromatographie – mit Beispielen aus Lebensmittel- und Umweltchemie", Springer, Berlin, p. 91

31. Chrompack International B. V., Middelburg, The Netherlands (1988), Chrompack News **15 (2)**: 5

32. CROSSLEY, A., HEYES, T. D., HUDSON, B. J. F. (1962) J. Am. Oil Chem. Soc. **39**: 9–14

33. NAWAR, W. W. (1969) J. Agric. Food Chem. **17**: 18–21

34. KITAMURA, K. (1971) Bull. Chem. Soc. Jpn. **44**: 1606–1609

35. NICHOLS, P. C., HOLMAN, R. T. (1972) Lipids **7**: 773–779

36. LIEN, Y. C., NAWAR, W. W. (1973) J. Am. Oil Chem. Soc. **50**: 76–78

37. BARR HIGMAN, E., SCHMELTZ, J., HIGMANN, H. C., CHORTYK, O. T. (1973) J. Agric. Food Chem. **21**: 202–204

38. DRAWERT, F., BECK, B. (1974) Z. Lebensm. Unters. Forsch. **155**: 1–9

39. ALENCAR, J. W., ALVES, P. B., CRAVEIRO, A. A. (1983) J. Agric. Food Chem. **31**: 1268–1270

40. HARDT, R. (1987) Thesis D83/FB 13 No. 225, Technische Universität Berlin, F.R.G.

41. BINNEMANN, P., JAHR, D. (1979) Fresenius Z. Anal. Chem. **297**: 341–356

42. BARFORD, R. A., MAGIDMAN, P. (1985) in "Modern Practice of Gas Chromatography" (Grob, R. L., Ed.), Second Edition, John Wiley & Sons, New York, Chichester, Brisbane, Toronto, Singapore (1985), p. 561

43. MELCHERT, H.-U. (1975) Dtsch. Lebensm. Rundsch. **71**: 400–402

44. ict-Handelsgesellschaft m.b.H., Frankfurt, F.R.G. (1987) "J&W Scientific General Catalog 1987/ 88", p. 29

45. Atlas of Chromatograms, GC59, in Ettre, L. S. (Atlas Consultant) (1988) J. Chromatogr. Sci. **26**: 190

46. Chrompack International B. V., Middelburg, The Netherlands (1987), Chrompack News **14 (1)**: 5

47. SIMMONDS, P. G., MEDLEY, E. E., RATCLIFF Jr., M. A., SHULMAN, G. P. (1972) Anal. Chem. **44**: 2060–2066

48. LIEN, Y. C., NAWAR, W. W. (1974) J. Food Sci. **39**: 911–913

49. LIEN, Y. C., NAWAR, W. W. (1974) J. Food Sci. **39**: 914–916

50. SHULMAN, G. P., SIMMONDS, P. G. (1968) Chem. Commun., pp. 1040–1042

51. BREITBART, D. J., NAWAR, W. W. (1979) J. Agric. Food Chem. **27**: 511–514

52. PICCAGLIA, R., GALETTI, G. C. (1988) J. Sci. Food Agric. **45**: 203–213

53. LOW, N. H., SPORNS, P. (1988) J. Food Sci. **53**: 558–561

54. GUERRANT, G. O., MOSS, C. W. (1984) Anal. Chem. **56**: 633–638

55. SCHERZ, H., MERGENTHALER, E. (1980) Z. Lebensm. Unters. Forsch. **170**: 280– 286

56. MERGENTHALER, E., SCHERZ, H. (1978) Z. Lebensm. Unters. Forsch. **166**: 225–227

57. PREUSS, A., THIER, H.-P. (1982) Z. Lebensm. Unters. Forsch. **175**: 93–100

58. PREUSS, A., THIER, H.-P. (1983) Z. Lebensm. Unters. Forsch. **176**: 5–11

59. HEYNS, K., STUTE, R., PAULSEN, H. (1966) Carbohydr. Res. **2**: 132–149

60. HEYNS, K., KLIER, M. (1968) Carbohydr. Res. **6**: 436–448

61. JOHNSON, R. R., ALFORD, E. D., KINZER, G. W. (1969) J. Agric. Food Chem. **17**: 22–24

62. PREY, V., EICHBERGER, W., GRUBER, H. (1977) Stärke **29**: 60–65

63. BALTES, W., SCHMAHL, H.-J. (1978) Z. Lebensm. Unters. Forsch. **167**: 69–77

64. VAN DER KAADEN, N, A., HAVERKAMP, J., BOON, J. J., DE LEEUW, J. W. (1983) J. Anal. Appl. Pyrolysis **5**: 199–220

65. SJÖBERG, A.-M., PYYSALO, H. (1985) J. Chromatogr. **319**: 90–98

66. HELLEUR, R. J. (1987) J. Anal. Appl. Pyrolysis **11**: 297–311

67. International Technical Caramel Association, Washington, DC, "Caramel Colors-Classification and Specification System"

68. HARDT, R., BALTES, W. (1987) Z. Lebensm. Unters. Forsch. **185**: 275–280

69. HARDT, R., BALTES, W. (1989) J. Anal. Appl. Pyrolysis, **15**: 159–165

70. HARDT, R., BALTES, W. (1988) J. Anal. Appl. Pyrolysis **13**: 191–198

71. DROSS, A., HARDT, R., BALTES, W. (1987) Fresenius Z. Anal. Chem. **328**: 495–498

72. HARDT, R., TSCHIERSKY, H., BALTES, W. (1988) Fresenius Z. Anal. Chem. **331**: 433–434

73. SANDRA, P., DAVID, F. (1988) in c.f. [1] (Sandra, P., Ed.), p. 300

74. MCKNIGHT HALKET, J., SCHULTEN, H.-R. (1988) Z. Lebensm. Unters. Forsch. **186**: 201–212

III Analysis of Residues and Contaminants

III.1 Pesticide Analysis

C. Fürst

1 Introduction

Pesticides are chemically and biologically active substances directed against animal-, vegetable- and microbiological pests. They are not only used for pests of cultivated plants, but also against those pests, which directly threaten the health of man and animals. Moreover, pesticides are employed against rodents, snails, household- and storage pests.

Pesticides mainly function to increase the yields in the production of agricultural goods and to improve their quality. In addition, they are of great health protection importance, especially in tropical and subtropical areas where they are applied to kill mosquitoes which carry malaria. Hence, pesticides not only help to augment food production around the world, but also save thousands of human lives.

Unfortunately, some pesticides do not break down very quickly and remain in the ecosystem for a long time. This is especially true for some lipophilic insecticides, like DDT and HCH, which over a period of years have accumulated in food chains. Although banned in North America and Europe in the 1970s, these compounds can still be determined in many food samples including human milk.

Newer and more effective pesticides break down more quickly reducing the chance of possible accumulation in organisms. Albeit ecologically more safe, these newer pesticides may still lead to residues in foods, especially if the withdrawal period between application and harvest has not been adhered to or the applied pesticide concentration was initially too high.

It is estimated that 500 to 600 different pesticides are used world-wide [1]. Depending on their biological activity and application, they may be classified into the following important classes:

- insecticides
- fungicides
- herbicides
- acaricides
- nematocides
- rodenticides
- fumigants
- plant regulators

Besides these pesticides, residues of other compounds have gained public interest in the past few years. This is for example the case for halogenated cleansing solvents, like tetrachloroethylene or halogenated carboxylic acids, and the increasing group of various environmental contaminants.

It is obvious that the huge number of potential residues makes analysis more difficult because in most cases the history of the sample to be analyzed is not known to the analyst.

2 Extraction

To effectively control residues of pesticides below maximum residue limits or to check for substances which are banned, it is necessary to use multiresidue procedures instead of methods which are only specific for a single compound. A simplified scheme of such multiresidue procedures as used for analysis of pesticides and also for the determination of veterinary drugs in foodstuffs of animal origin is depicted in Figure III.1.–1 [1].

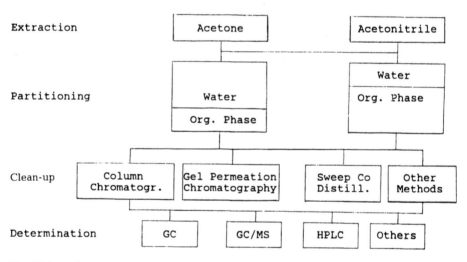

Fig. III.1.–1 Simplified scheme for multiresidue procedures [1]

Some substances undergo metabolism or are bound to tissues or cell walls. This is especially important for the analysis of pharmacologically active substances, which may have been transformed into glucuronides and sulfates or show a strong binding to proteins [2]. For the hydrolysis of these conjugates and bound residues and the release of the genuine active compounds, mostly enzymatic methods, which are milder than acid or alkali treatments, are used.

Compared with samples of animal origin, hydrolysis of metabolites in vegetables or fruits is normally of less importance. Only those pesticides, which penetrate into the interior of the cells are transformed to a certain degree [1]. This was reported e.g. for the widely used herbicides 2.4-dichlorophenoxyacetic acid, linuron, diuron and the fungicide pentachlorophenol [3–5].

When choosing a solvent for extraction, some important aspects should be considered.

– solubility of the active substance of interest
– solubility of naturally occurring compounds (water, fat, proteins and carbohydrates)
– behaviour of the foodstuff towards the solvent
– toxicity

Common solvents for extraction in order of increasing polarity are:

dichloromethane < acetone < ethyl acetate < acetonitrile < ethanol < methanol

For residue analysis methods in particular, it has proved very worthwhile to extract active substances by homogenizing the sample with polar organic solvents. For the analysis of acidic or basic compounds it is important to adjust the pH to a value where the substances of interest are largely non-dissociated prior to analysis.

Amphoteric compounds show the lowest water solubility at the isoelectric point and should be extracted with an organic solvent at the corresponding pH-value.

In principle, it is possible to perform residue extraction of pesticides with water-soluble (acetone, acetonitrile, methanol) or water-insoluble solvents (ethyl acetate, dichloromethane, diethyl ether). The latter have the advantage that the majority of sample ingredients remain in the water phase, which facilitates further clean-up steps. The disadvantage of these solvents is that, because of their lipophilic properties, they cannot penetrate into the interior of cells where hydrophilic conditions are predominant. This leads to poorer recoveries of biologically adsorbed residues bound to proteins.

During the extraction of food samples the possible formation of intractable emulsions may become very problematic. This can clearly be demonstrated by homogenizing liver with dichloromethane, which results in a creamy consistency making further clean-up impossible [6].

Toxicity is an additional criterion, which should be considered when choosing a solvent. For health reasons, solvents, like chloroform or benzene should be avoided.

From the above considerations, it follows that there exists no ideal solvent. But two solvents, acetone and acetonitrile, come very close to being ideal and thus are widely used for residue analysis. Due to their polar properties and their excellent water solubility, they are able to penetrate into the interior of tissues usually leading to high recoveries for active substances with only minor co-extraction of unwanted plant or tissue materials.

Because both solvents denature proteins and only marginally dissolve carbohydrates, this kind of extraction is already combined with a first clean-up step.

However, when working with acetonitrile all procedures should be performed in a fume hood because this solvent might also be harmful to one's health.

Using acetone or acetonitrile, water is also extracted from the sample. This may easily be removed by saturation with sodium sulfate followed by addition of dichloromethane. In this way high recoveries can be obtained even for active substances, which have high water-solubility [7, 8].

When analyzing lipophilic compounds, which are mainly stored in fatty tissues, the first step normally consists in extraction of the lipid together with the substances of interest, followed by a clean-up step to remove the fat. This applies to the determination of organochlorine pesticides, such as DDT, HCB and HCH as well as for most halogenated environmental pollutants.

3 Clean-up

The choice of a clean-up procedure is dependent on the type of detection system to be used for the analytical determination. Normally investigations employing combined gas chromatography-mass spectrometry (GC-MS) require less extensive clean-up steps than analyses with a gas chromatograph equipped with an electron capture detector.

The following clean-up procedures are mostly used in residue analysis:
- adsorption column chromatography
- gel permeation chromatography
- sweep co-distillation and
- liquid/liquid-partitioning

3.1 Adsorption Column Chromatography

In the scientific literature a multitude of clean-up procedures based on adsorption column chromatography have been described. THIER and FREHSE give an actual comprehensive review on the various feasibilities with respect to residue analysis [1].

Common adsorbants are Florisil, alumina, silica gel in different modifications and charcoal.

Because of the introduction of mini cartridges, like Sep-Pak, Bond-Elut or Baker cartouches, a scaling down of purification steps has been possible in the past few years. Due to the minimal space required and the extensive accessories commercially available, these cartridges also offer the advantage that large sample sets can be worked up at the same time. This explains inter alia the increasing importance of these systems especially in water analysis.

Using adsorbants, one should always be aware of the possibility of different separation properties depending on the lot. Because most separations are carried out within a narrow range of conditions, it is absolutely necessary to recheck the elution profile with standard compounds when working with a new lot. It should also be noted that conditions optimized for a material from one manufacturer in most cases cannot be adopted without qualification for a product from a second manufacturer although the adsorbants seem to be the same.

Despite all the restrictions mentioned, adsorption column chromatography is the clean-up procedure most widely used in residue analysis. This is also because it is possible using suitable conditions not only to remove the majority of unwanted co-extractives but also to fractionate the compounds of interest into several eluates of different polarity, which facilitates the subsequent analytical determination.

SPECHT and TILLKES describe a multiresidue procedure for the analysis of lipid- and water-soluble pesticides in food- and feeding stuffs of vegetable and animal origin [8]. The key step of this method is the fractionation of a crude extract on a mini silica gel column containing only 1 gram adsorbant. Using various solvents of increasing polarity, one obtains up to seven eluates of 10 ml each. In this way, for example, polychlorinated and polybrominated biphenyls are separated from most of organochlorine pesticides thus reducing the possibility of false positives and making quantification easier. Elution data for more than 400 active substances have been reported so far [8]. Another major advantage of this procedure is that it provides efficient identification and determination even of those compounds which cannot be separated on a capillary column, provided they show a different elution profile on the mini silica gel column.

3.2 Gel Permeation Chromatography

Gel permeation chromatography (GPC) is a very mild and universal clean-up procedure sometimes also described as size-exclusion chromatography. GPC mainly separates according to molecular weight in which exclusion size is dependent on the gel material. Clean-up procedures using GPC have been reported for the analysis of pesticides [7−10], environmental contaminants [11−13] as well as for residues of pharmaceutically active substances in foodstuffs [6, 14, 15].

Whereas Bio-Beads S-X3 (Bio-Rad), a copolymer of styrene and 3 % divinyl benzene with a mesh size of 200−400, is mostly used for pesticide and environmental analysis, Sephadex gels are preferred for the determination of pharmaceutically active substances.

A major advantage of GPC is the possibility for automation. The GPC AUTO PREP 1001 (Analytical Biochemistry Laboratories, Inc.) which is widely used in residue analysis is equipped with 23 sample loops each containing a volume of exactly 5 ml. With this instrument one may clean-up 23 samples unattended overnight.

It had previously been considered as a disadvantage that at least 7–8 ml of sample have to be injected onto this system to overcome the dead volume between injection valve and sample loop. This involves a considerable loss of extract potentially leading to a poorer sensitivity because the sample loops can only take up 5 ml. This problem can be overcome by using an injection technique according to NORSTRÖM et al. [12]. Actually, this technique requires several sample loops per extract, which reduces the total number of specimens automatically cleaned up overnight. But this restriction is of minor importance in particular for the analysis of environmental pollutants in the ppt and ppq range where it is absolutely necessary to minimize potential losses during the clean-up in order to achieve a sensitive analytical determination.

To establish the elution times of the compounds from the GPC-column, elution profiles for the active substances as well as for extracts of foodstuffs to be analyzed must be determined. Once the elution profile of the active substances is known, fractionation using a very narrow volume range can be carried out in order to remove as many co-extractives as possible.

After changing columns, elution profiles have to be rechecked because swelling of the gel is highly dependent on the lot. But even appyling crude extracts, a GPC-column normally survives several thousand analyses.

3.3 Sweep Co-Distillation

Sweep co-distillation (SCD) is a relatively fast and simple method suitable for a multitude of different compound classes [16–18]. It is mainly used for the residue analysis of vegetables.

The principle of this technique consists in first condensing an extract as a thin layer on an inert material in a specific column. The volatile compounds are then stripped off from the condensate either by injection of solvent or by addition of inert gas and collected in a cold trap. In most cases this technique yields very clean extracts because only volatile compounds are recovered.

The only restriction is that substances, which are volatile but degrade cannot be analyzed by means of sweep co-distillation.

Compared to the milder, but more time-consuming gel permeation chromatography, sweep co-distillation has its benefits when quick results and conclusions are required. This is especially true for regulatory pesticide analysis as part of import control.

3.4 Liquid/Liquid-Partitioning

Liquid/liquid-partitioning is based on the different solubility of compounds in two non-miscible solvents. For a better partitioning, the solvents may be saturated with the corresponding partner.

Separatory funnels used for liquid/liquid-partitioning in residue analysis should be equipped with inert stopcocks, such as teflon or PTFE. If only glass stopcocks are available, one must not lubricate, but only moisten them with a droplet of water.

Because liquid/liquid-partitioning is very labor- and sometimes time-consuming, especially when emulsions are formed, their importance in residue analysis decreases in favor of the above mentioned procedures.

Exceptions are, however, the analysis of acidic and basic substances, which can be easily separated from the matrix by means of acid/alkaline partitioning. This was reported for the fungicide pentachlorophenol and chlorinated phenoxy carboxylic acids widely used as herbicides [19, 20]. In both cases, the separation of the active substances from the majority of the matrix was performed using a borate buffer pH 10.

Partitioning an extract between an organic layer, like hexane and sulfuric acid is a special case of liquid/liquid-partitioning [21, 22]. This is a highly effective procedure to degrade and remove lipids from PCBs and other lipophilic compounds, such as polychlorinated dibenzodioxins and dibenzofurans.

In contrast, such pesticides as dieldrin and methoxychlor are not stable under these conditions and therefore are also degraded.

Summarizing, it can be stated that for the clean-up of food extracts a variety of feasibilities are reported in the scientific literature. Depending on the type of sample and active substance to be analyzed, these procedures should be combined like modules to give extracts as clean as necessary or possible.

Robots constitute the latest development in sample extraction and clean-up. Recently MES reported on the use of a robot for analysis of organochlorine pesticides and polychlorinated biphenyls (PCBs) in milk [23]. In this case, the robot performs the complete extraction and clean-up procedure, from weighing samples through fractionation of organochlorine pesticides from PCBs by use of adsorbant silica column chromatography. The final extract can directly be applied to gas chromatographic analysis. Compared with manual extraction and clean-up, the precision of the procedure was pretty much the same. This approach will certainly be more ubiquitious in the future because robots can operate unattended and thus make a high sample throughput possible.

4   Detection

Depending on the type of active substance to be analyzed, various detectors are available. For residue analysis, the following detectors are most important:
– electron capture detector (ECD)
– nitrogen-phosphorus detector (NPD, TID, PND, N-FID)
– mass selective detector (MS, MSD, Ion Trap)

Besides these, further detection systems, such as flame photometric detectors (FPD) for selective identification and determination of sulfur- and phosphorus-containing substances as well as electroconductivity detectors (e.g. Hall detector) for the determination of nitrogen-, sulfur- or halogen-containing compounds are occasionally used.

4.1 Electron Capture Detector

Introduction of the electron capture detector (ECD) in the early sixties substantially contributed to the importance of gas chromatography in residue analysis [1, 24–26]. Because of an extreme sensitivity for halogenated compounds, it has been classically used for the determination of organochlorine pesticides and polyhalogenated pollutants.

However, the ECD response is not only dependent on the type and number of halogen atoms, but also on their position in the molecule. This may probably lead to difficulties when quantitating complex mixtures like PCBs, because the various isomers of a given group can not be calculated by using the same response factor. For this, a precise quantitation can only be achieved when the actual substance of interest is available as a reference standard, and therefore a calibration can be performed.

In case of polychlorinated and polybrominated biphenyls (PCBs, PBBs), this is almost impossible because only a very limited number of distinct isomers is commercially available. With respect to these two classes of environmental pollutants, the difficulties just mentioned, may be overcome by using a flame ionization detector for an intermediate quantification. This will be explained in detail in the chapter on contaminants (III. 2).

Analyzing samples of vegetable origin, compounds to be determined may occasionally be superimposed by matrix compounds of the plant, which are normally present at a large surplus. This is especially true for esters and sulfur-containing compounds typical of onions and garlic, which, due to their high concentration, also give a signal on the ECD.

Another group of compounds showing strong ECD signals are phthalates widely used as plasticizers. In these cases, further clean-up steps or other detection systems such as combined gas chromatography-mass spectrometry (GC-MS) should be used to differentiate between active substances and matrix compounds.

Despite this limited selectivity, the ECD is still the most important detector in residue analysis because it shows not only a high sensitivity for halogenated compounds, but also for nitro- and thiophosphate-containing phosphoric acid esters.

Further relevant uses for ECD are the analysis of halogenated solvents, disinfectants and cleansing agents. Examples are the determination of tetrachloroethylene in fatty food samples [27] and the analysis of halogenated carboxylic acids in beer and sparkling wine [28].

4.2   Nitrogen-Phosphorus Detector

Because of its high sensitivity for phosphorus- and nitrogen containing compounds, the nitrogen-phosphorus detector (NPD, TID, N-FID) is the detector of choice for the analysis of organophosphorus pesticides and triazine herbicides. Moreover, it is best suitable for the determination of nitrogen containing pharmaceutically active substances in foodstuffs. In contrast to the ECD, this detector belongs to the highly selective ones [24–26]. Although simpler constructed than the ECD, the nitrogen-phosphorus detector normally demands more "fine tuning" from the analyst when optimizing detector conditions.

Important criteria for an optimum sensitivity and a high selectivity are the temperature of the alkali salt bead, its distance from the nozzle and the ratio of hydrogen and air. The change of this ratio shifts selectivity and sensitivity towards nitrogen or phosphorus containing compounds.

Due to the high selectivity of nitrogen-phosphorus detectors, normally no extensive clean-up steps are necessary when analyzing food extracts.

However, despite the high selectivity and use of capillary columns, it should be mandatory to perform the chromatographic analysis on two capillaries of different polarity. This should be part of quality assurance in residue analysis.

4.3   Multidetection Systems

Information from residue analyses performed on electron capture and nitrogen-phosphorus detectors complement one another. Therefore the two detectors are occasionally used in parallel [29–31]. This can be achieved by splitting the carrier gas at the end of the capillary column and directing it simultaneously onto the two detectors or by mounting the NPD on top of the ECD. In the second case this is possible because the electron capture detector works without degradation of the compounds.

Figure III.1.–2 shows a schematic diagram of the multidetection system proposed by DELEU and COPIN [29]. In this case, an ECD is connected in series with a nitrogen specific detector (NPSD) and a sulfur specific detector is used in parallel. A $^1/_{32}"$ zero dead volume tee with special adapters is used for splitting of the effluent. Using this technique, one gains a high degree of information on the probable occurrence of various compound classes by only performing one injection.

4.4   Combined Gas Chromatography-Mass Spectrometry

In the past few years, combined gas chromatography-mass spectrometry (GC-MS) has become a powerful tool in residue analysis. This technique provides the analyst not only with full mass spectral information important for structure elucidation of unknown compounds, but also with high sensitivity and selectivity for selected ion monitoring (SIM) nor-

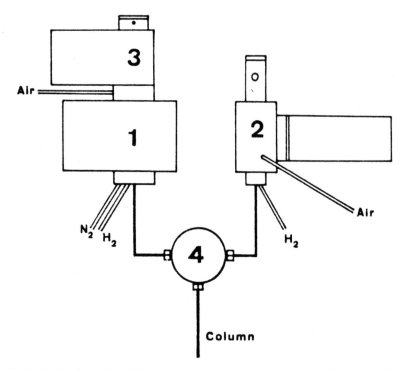

Fig. III.1.–2 Configuration of the multidetector system according to Deleu and Copin [29]
1 ECD; **2** SSD; **3** NPSD; **4** Dead volume coupling

mally employed in trace analysis. Moreover, current technology has led to a considerable price reduction. Application of menu driven and user-friendly software made operation easier. The best examples are the introduction of mass selective – [32] and ion trap detectors [33].

In principle, two different modes of GC-MS operation can be distinguished:

– Acquisition of full mass spectral data (Full Scan)

– Mass fragmentography (selected ion monitoring (SIM) or multiple ion detection (MID))

4.4.1 Full Scan

In the case where the identification of unknown peaks is needed, it is normally necessary to acquire complete mass spectra. Especially with respect to residue analysis, it can be often observed that, when using multiresidue procedures with ECD or NPD, signals are detected whose retention times do not correspond with those of reference compounds even if hundreds are available. In this case, GC-MS analysis leads to important information if the peak in question is an active substance or an interfering matrix compound, which because of a high concentration also gives a signal on the above mentioned detectors.

Besides the identification of unknown compounds, combined gas chromatography-mass spectrometry may also serve well for confirmation of results. Depending on the concentration of the compound in question and equipment available, this can be achieved either by mass fragmentography or acquisition of complete mass spectra. For example, if ionization is performed in the electron impact mode (EI), the conventional and most widely used ionization technique, analysis requires approximately 1–10 nanogram to record a complete mass spectrum. Hence, this mode can only be applied to compounds occurring at relatively high levels.

On the other hand, if results, which were obtained with highly sensitive detectors, like ECD and NPD, have to be confirmed, the concentration, which is sometimes only in the low picogram range is normally too small to acquire a complete mass spectrum in the EI-mode. In this case, confirmation of results may be performed using either mass fragmentography or with the aid of other ionization techniques, such as chemical ionization.

Chemical ionization belongs to the so-called soft ionization techniques [34]. During the analysis a constant stream of reagent gas, mostly methane or iso-butane, flows into the source of the mass spectrometer. Whereas electron impact ionization is normally performed at a source pressure of 10^{-6} to 10^{-7} torr, the addition of reagent gas results in a source pressure of approximately 10^{-4} torr, ideal for chemical ionization. Because of the large excess, the reagent gas is ionized at first, and in turn ionizes the analytes eluting from the column into the mass spectrometer. The resulting mass spectra in most cases show an intense molecular ion and only very little fragmentation. Using chemical ionization, it is often possible to determine the molecular weight of substances for which the molecular ion is not obtainable in the electron impact mode due to their instability under the latter ionizing conditions.

If compounds contain electron-withdrawing substituents, chemical ionization can result in the formation of negative ions caused by resonance electron capture (NCI). This phenomenon can be observed from the analysis of halogenated pesticides, which under suitable conditions yields high ion currents making a very sensitive NCI determination possible. For some compounds, only picogram amounts are needed to acquire a complete mass spectrum. In practice, special equipment is required for recording negative ions.

However, the major advantage of this technique is that compounds with a suitable structure can be determined with high selectivity and extreme sensitivity. The increasing importance of NCI-measurements in residue analysis can be attributed to the above mentioned facts.

4.4.2 Selected Ion Monitoring

With the aid of selected ion monitoring (SIM), often also called selected ion recording (SIR) or multiple ion detection (MID), quantification of trace residue levels can selectively be achieved by combined GC-MS. This technique is preferentially used when the substances to be determined are already known. For this purpose, the mass spectrometer is employed as a mass selective detector.

A prior knowledge of the mass spectra of substances to be analyzed is the basic requirement for SIM analyses. Characteristic ions are then chosen from the spectra. With regard to high sensitivity and selectivity, fragments showing intense relative abundances and ions with a high mass-to-charge ratio are of special interest. During the analytical run only a few selected fragments and not complete mass ranges are recorded. This technique leads to an enhanced sensitivity because the scan time for each fragment is considerably longer. Moreover, clean chromatograms are normally obtained because only those compounds are determined, which contain at least one of the specific fragments chosen for analysis.

5 Specific Applications of Capillary Gas Chromatography

Whereas in former years many packed columns were necessary when dealing with residue analysis, more than 90 % of the problems related to residues of pesticides and contaminants, which are accessible by gas chromatographic analysis, can be investigated using only three columns of different polarity today. For almost all compounds it has proven feasible to perform the analysis on a nonpolar as well as on a medium polar column because of their relatively high maximum operation temperature. However, more polar compounds, like phenols or alcohols, may require the use of polar columns.

The introduction of cross-linked and chemically bonded fused silica columns has not only facilitated the handling of capillary columns, but also improved their thermostability and reduced the ability of column bleed. Hence, these column types are preferred in residue analysis. In any case, it should be noted that the life-time of columns can be prolonged by placing water- and oxygen traps in the carrier gas lines.

High numbers of routine analyses make automation necessary. Modern laboratory data systems not only control automatic liquid samplers, but also process the raw data from integrators and store the results for future calculations or statistics. For this, gas chromatographs can be operated unattended even overnight, a basic requirement for a high sample throughput.

The following examples illustrate the use of capillary columns in combination with different detectors in residue analysis.

No emphasis has been given on a systematic classification of active substances. This is described in detail elsewhere [1, 35–37]. The reader is also referred to the papers published by the Food and Agriculture Organization (FAO) of the United Nations [38]. These papers, named "Pesticide Residues in Food" comprise the results of the annual joint meetings of the FAO panel of experts on pesticide residues in food and the environment and the WHO expert group on pesticide residues with respect to all aspects related with pesticides.

Comprehensive reviews, progress reports and archival documentations are also included in the "Reviews of Environmental Contamination and Toxicology", the continuation of the former "Residue Reviews" [39].

5.1 Insecticides

Organochlorine pesticides (OCP) comprise lipophilic and persistent compounds, such as DDT, hexachlorocyclohexane (HCH), aldrin, dieldrin and heptachlor. These active substances belong to the early generation of synthetic organochlorine compounds, which played a significant role as insecticides in pest control over the past 30–50 years. Due to their lipophilic properties and high persistency, these pesticides bioaccumulate in food chains and therefore have been banned in almost all countries of the western world. Nevertheless, residues and metabolites of these compounds can still be determined in various food samples, and especially in human milk.

Residue analysis of organochlorine pesticides is mostly performed by capillary gas chromatograhy on two columns of different polarity with electron capture detection (ECD).

Figure III.1.–3 shows the gas chromatographic separation of a human milk extract on a SE-30/52 and an OV-1701 column, respectively [40]. The organochlorine pesticides were separated from PCBs according to the multiresidue procedure of SPECHT and TILLKES [7, 8]. The effluent stream from the SE-30/52 column was split and directed to an ECD and TID, which were placed in parallel. In this way, one obtains important structure-specific information from the compounds in addition to their retention times.

It can be seen that oxychlordane and cis-heptachlorepoxide are only partly resolved on the SE-30/52 column. Their simultaneous presence could, however, be confirmed by analysis on the OV-1701 column, which unequivocally separates both compounds. At the end of the analytical run, a computer program integrates the signals and perfoms the identification of the compounds by comparing their measured retention times with identification times of corresponding reference standards stored in a calibration file [40]. Finally, a compressed report including the name of the compound, its actual retention time (RT) and identification time (ID-TM) as well as the concentration measured on both columns is printed out (Fig. III.1.–4).

This mode of operation represents a valuable aid for the analyst. However, it does not replace his expertise in interpreting the results.

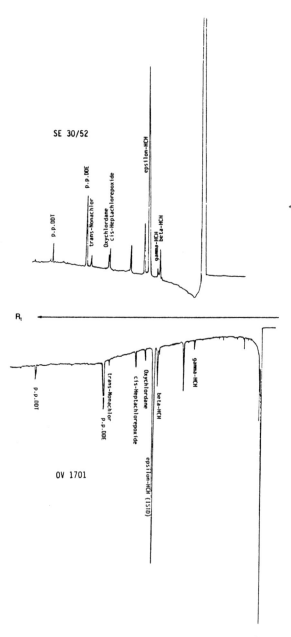

Fig. III.1.–3 Gas chromatographic separation of a human milk extract on a 50 m SE-30/52 and OV-1701 column, respectively [40]

```
-----------------------------------------------------------------------
RESULT FILE:/DATA/LOOP/RESULT/MB023.RES
RESULT FILE:/DATA/LOOP/RESULT/OB023.RES

RT(30/52)  ID-TM   MG/KG   RT(1701)  ID-TM   MG/KG   NAME
=======================================================================
  11.70    11.72   .093     12.33    12.35   .100    BETA-HCH
  12.03    12.04   .011      8.46     8.48   .013    GAMMA-HCH
  13.04   -13.0    .481     12.70   -12.7    .481    EPSILON-HCH (ISTD)
  18.21    18.23   .076     14.44    14.45   .077    CIS-HEPTACHLOREPOXIDE
  18.33    18.31   .027     13.48    13.49   .025    OXYCHLORDANE
  20.52    20.49   .046     17.18    17.17   .047    TRANS-NONACHLOR
  21.16    21.11   .104     17.70    17.74   .108    P.P.DDE
  25.68    25.72   .047     24.73    24.74   .052    P.P.DDT
  32.04    31.95   .599     30.84    30.87   .636    DIETHYLHEXYLPHTHALAT
```

Fig. III.1.–4 Compressed report of a human milk analysis for residues of organochlorine pesticides [40]

Identification of compounds based on their absolute retention times may become problematic if columns are aging or have to be shortened due to deposits of nonvolatile compounds at the column inlet. For this reason, several attempts have been made to develop specific index standards that offer similar feasibilities in residue analysis as n-alkanes, which serve as a basis for the KOVATS-index system for FID measurements [41, 42].

Recently MANNINEN et al. reported on the synthesis of a homologous series of alkylbis-(trifluoromethyl)phosphine sulfides (M-series), which have the specific label atoms at one end of a hydrocarbon chain [43].

The structure of these compounds is shown in Figure III.1.–5. Because these substances are thermally stable and give responses on flame ionization-, nitrogen-phosphorus-, flame photometric- and electron capture detectors, they ideally serve as universal retention index standards. Application of the multidetector retention index standards (MDRI) to the residue analysis of a sweet pepper sample on a dual channel GC system is demonstrated in Figure III.1.–6 [43].

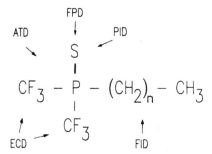

Fig. III.1.–5 Structure of alkylbis(trifluoromethyl)phosphine sulfide

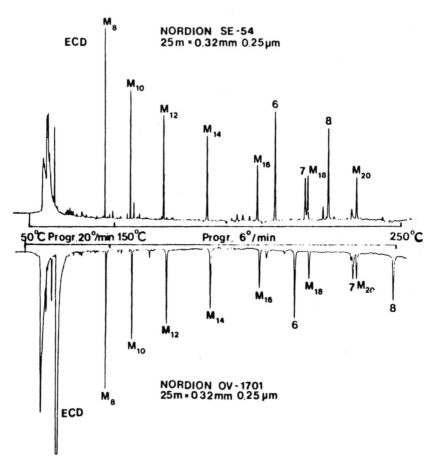

Fig. III.1.–6 Gas chromatographic separation of a sweet pepper sample on a dual channel GC system [43]

M_8–M_{20} : Index standards with even carbon numbers C_8–C_{20}

6 = alpha endosulfan; **7** = beta endosulfan;

8 = endosulfan sulfate

Peaks marked M_8 through M_{20} are the index standards with even carbon numbers
(C_8–C_{20}) and peaks 6, 7 and 8 represent three pesticides (alpha-endosulfan, beta-
endosulfan and endosulfan sulfate), which were determined in this sample. Identification
of the compounds was achieved by applying the cubic spline retention index method
[44].

Complete systems comprising dual channel GCs and a data station which is capable of controlling an autosampler as well as identifying compounds based on the multidetection retention index standards are now commercially available [45].

An elegant method of analyzing one extract on two capillary columns of different polarity consists in application of two-dimensional gas chromatography [31, 46–50].

STAN et al. reported on the analysis of about 100 halogenated and organophosphorus pesticides [31] and 57 organophosphorus pesticide residues in food [46], using two-dimensional gas chromatography with pneumatic column switching according to DEANS [51, 52].

The principle of this technique, also called "live chromatography", consists in changing the direction of the effluent flow between columns by means of make-up gas generated by activation of external valves. The system works with slight pressure differences adjusted by two make-up gas lines between the two ends of a T-piece. Compounds separated on the first column can either be directed to the detector or further analyzed with a second column.

Application of this technique to the analysis of several organophosphorus pesticides is depicted in Figure III.1.–7.

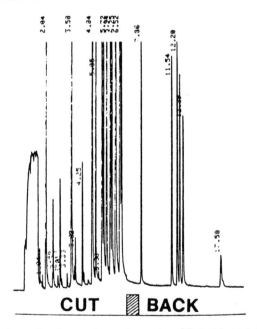

Fig. III.1.–7 Two-dimensional gas chromatography (SP 2100 and OV-225) of organo-phosphorus pesticides [46]

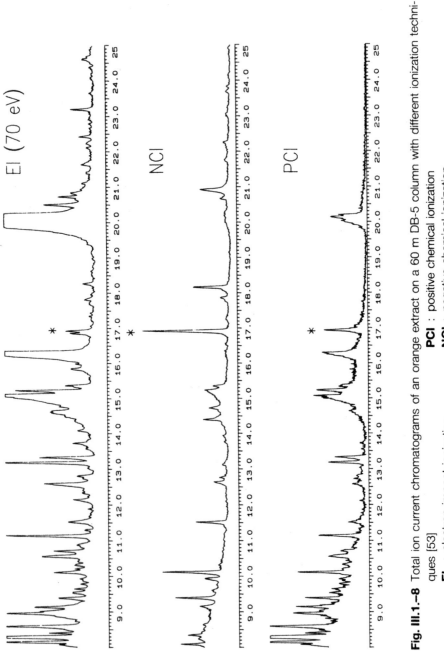

Fig. III.1.–8 Total ion current chromatograms of an orange extract on a 60 m DB-5 column with different ionization techniques [53]

PCI : positive chemical ionization

NCI : negative chemical ionization

EI : electron impact ionization

It illustrates the identification of six organophosphorus pesticides, which are not completely separated on the first column (SP 2100). Hence, this group was cut out from the mixture by directing the effluent to the second column of different polarity (OV-225). The cut procedure is indicated by the striped box. After the last peak in question had been transferred to the second column, the backflush was activated in order to prevent co-elution with late eluting substances from the first column. Identification of the compounds is finally achieved by comparing their total retention times with those of standard compounds.

If pesticides exceed regulatory maximum residue limits, confirmation of the results should be performed using a different detection system such as GC-MS.

Figure III.1.–8 shows total ion chromatograms of an orange extract obtained from three GC-MS runs with different ionization modes [53]. In this case, ECD and NPD analysis revealed the presence of the phosphoric acid ester, chlorpyrifos. GC-MS-analysis was employed to ensure that the peak really is the above mentioned insecticide and not a matrix compound, which due to an excess amount also generates a signal on the ECD. The peak of interest is marked by an asterisk. It is interesting to compare the height of this peak as a function of the ionization technique applied. It can be clearly seen that the peak in question shows the highest response in the NCI chromatogram. From this, one might already deduce that this signal most likely represents a halogenated compound.

A library search of the mass spectrum obtained in EI mode confirmed the presence of chlorpyrifos. The good concurrence of the mass spectra of unknown peak (A) and the best match gained from the library search (B) is illustrated in Figure III.1.–9.

The availability of mass spectral libraries is very useful. Modern GC-MS systems contain libraries with up to 80 000 compounds and perform a computer based search of an unknown mass spectrum in less than one minute. Unfortunately, to date only mass spectral libraries acquired in EI mode are commercially available because PCI- and NCI spectra are basically dependent on the type of reagent gas and the MS conditions, like source temperature and source pressure. For this, PCI- and NCI spectral libraries have to be created by the user himself.

Although GC-MS libraries contain thousands of spectra, occasionally the compound of interest is not included. In this case, interpretation of the mass spectrum normally has to be done manually, albeit there already exists an automatic approach for this purpose.

The self-training interpretive and retrieval system (STIRS) for interpretation of unknown mass spectra is a package of computer programs developed by MCLAFFERTY and co-workers that supports the mass spectrometrist in the analysis of mass spectra of unknown compounds [54, 55].

For manual interpretation, it is helpful to perform data acquisition by means of different ionization techniques. Figure III.1.-10 shows the GC-MS analysis of a tomato extract in which an unknown signal was detected in both an ECD and NPD run [53].

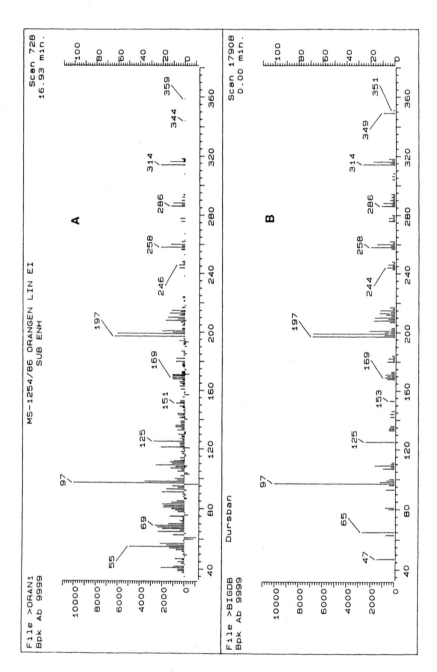

Fig. III.1.–9 Result of a MS library search
A : Mass spectrum of unknown peak
B : Mass spectrum of best match from library search
(Dursban is the trade name of chlorpyrifos)

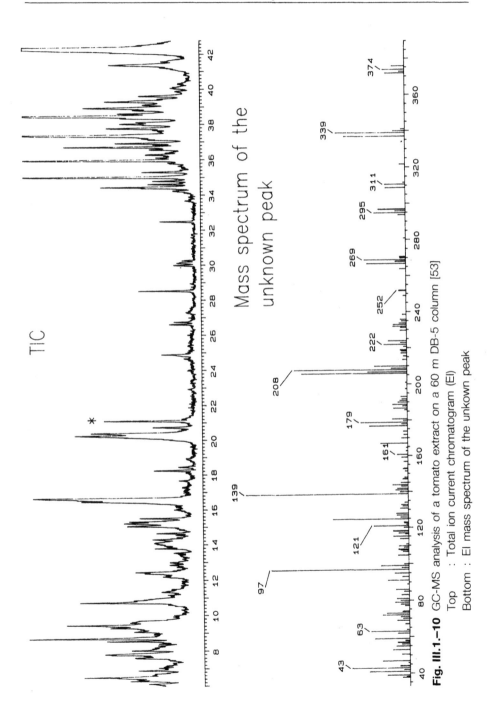

Fig. III.1.–10 GC-MS analysis of a tomato extract on a 60 m DB-5 column [53]
Top   : Total ion current chromatogram (EI)
Bottom : EI mass spectrum of the unkown peak

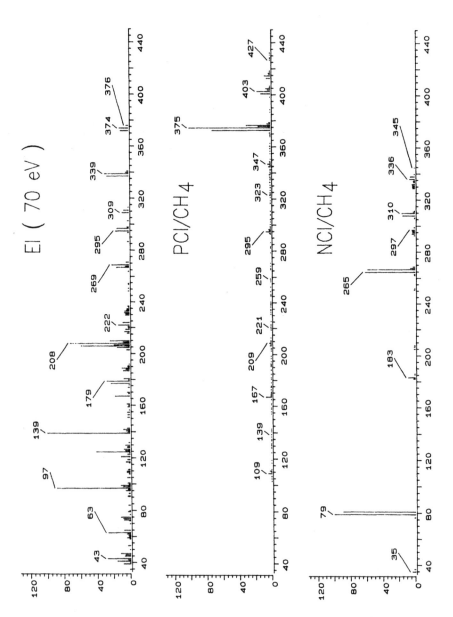

Fig. III.1.–11 Mass spectrum of an unknown peak in dependence on the ionization mode [53]

The upper chromatogram represents the total ion current recorded in electron impact mode (EI). Whereas NCI runs normally look like ECD chromatograms, EI runs are more or less similar to analyses using a flame ionization detector (FID). Hence, a lot of substances, mainly flavor compounds, which do not give a signal on an ECD are also detected in the upper chromatogram. This makes GC-MS analysis in EI mode more difficult because the ratio of the peak of interest and surrounding peaks differs considerably in the two detection modes. For example, a polyhalogenated substance giving a strong signal on an ECD may only reveal a weak peak in a GC-MS-EI run. A flavor compound like a terpene is not detected from an ECD but generates a large signal in an EI run. It should also be considered that retention times normally cannot be compared because most GC-MS systems demand helium as carrier gas whereas hydrogen and nitrogen are preferred for other GC investigations. For this, use of relative retention indices facilitates identifying the peak in question amongst a bunch of other signals.

The mass spectrum of the peak marked by an asterisk is depicted in the lower part of Figure III.1.–10. A library search revealed no positive result. Hence, the PCI and NCI spectra were also acquired. These are shown in Figure III.1.–11.

The NCI spectrum clearly shows that one or more chlorine and bromine atoms (m/z 35, 37, 79 and 81) must be part of the compound.

Positive chemical ionization (PCI) reveals that due to the protonated "quasimolecular ion" at m/z 373, the molecular weight is most likely 372. Moreover, the characteristic isotope pattern indicates that the compound contains only one chlorine and bromine atom, respectively. Higher fragments, which are caused by addition products of the reagent gas, are very important for identifying the molecular ion.

From the EI spectrum one can deduce inter alia the presence of a phosphorothionate group (m/z 125).

Based on the combined information gained from the three mass spectra, this compound was finally identified as the phosphoric acid ester prophenofos, an insecticide, which is not registered in the Federal Republic of Germany.

5.2 Herbicides

Herbicides are chemically active substances, which control unwanted plants that infest agricultural crops. This group of pesticides comprises a variety of different compound classes such as triazines, urea derivatives, carbamates, thiocarbamates, amide derivatives and many others [1]. The most important herbicides are, however, chlorinated phenoxycarboxylic acids and their derivatives. These compounds have found wide-spread use in pest control because they promote the development and production of cultivated plants by inhibiting weeds, which are mostly dicotyledons.

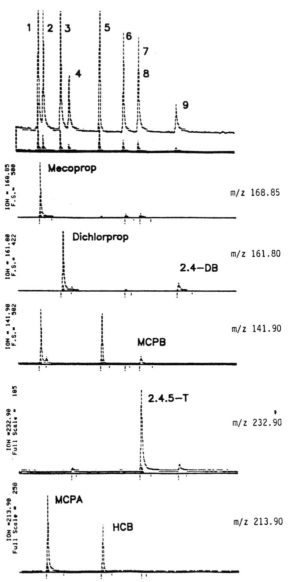

Fig. III.1.–12 Mass fragmentograms of a standard mixture containing eight chlorinated phenoxycarboxylic acids as methyl esters and HCB as internal standard, separated on a 30 m SE-30 glass capillary column [20]

1 = Mecoprop	**2** = MCPA	**7** = MCPB	**8** = 2,4,5 – T
3 = Dichlorprop	**4** = 2,4 – D	**9** = 2,4 – DB	
5 = HCB (ISTD)	**6** = Fenoprop		

Chlorinated phenoxycarboxylic acids are relatively persistent compared with other herbicides. Thus, use of these chemicals most likely results in residues, especially in grain, grain products, wild mushrooms and -berries.

Because chlorinated phenoxycarboxylic acids are polar and less volatile, they have to be derivatized prior to gas chromatographic analysis. Methylation results in a poor ECD response for the monochloro-compounds MCPA, mecoprop and MCPB making ECD analyses at residue levels almost impossible. Enhanced sensitivity can be achieved by derivatization with pentafluorobenzylbromide, 2, 2, 2,-trichloroethanol, 2, 2, 2,-trifluoroethanol and similar halogen containing compounds [56]. These reagents cover, however, also other acidic co-extractives, which generally results in a high number of interfering signals in GC/ECD analysis, making an unambiguous determination of active substances very difficult.

The problem just mentioned can be overcome by employing a mass spectrometer instead of an ECD for the detection of chlorinated phenoxycarboxylic acids. Because of the high selectivity and sensitivity of mass selective detection, an introduction of halogen containing groups is superfluous. Moreover, purification of samples can be reduced to a minimum [20].

Figure III.1.–12 shows mass fragmentograms of a standard mixture, which contains the most important chlorinated phenoxycarboxylic acids as their methyl esters [20]. Separation of the compounds was carried out on a 30 m SE-30 glass capillary column. MS analysis was performed in the SIM mode with electron impact ionization. One specific ion for each active substance was chosen for screening runs. HCB served as an internal standard. The topmost chromatogram represents the sum of the ion current measured for all prechosen mass fragments. The other traces were subsequently reconstructed by choice of specific mass fragments for each compound. Although MCPB and 2,4,5-T co-elute completely, a definite identification and quantification of both herbicides is possible due to their different mass fragments, even if they are present simultaneously.

Fig. III.1.–13 shows mass fragmentograms of a wheat specimen in which dichlorprop is present at a concentration level of 0.01 mg/kg (ppm) [20]. Five specific ions were measured. Due to the fact that besides the retention time also the relative ratio of their peak areas was in good agreement with those of the reference standard, the presence of dichlorprop can be regarded as confirmed.

Triazines, like atrazine and simazine, represent another class of herbicides with increasing importance, especially in corn cultivation.

Due to their wide-spread use in some countries, these substances have already found their way into drinking water. Because of the presence of nitrogen atoms in their molecules, a NPD is the detector of choice for the residue analysis of triazine herbicides. In most cases, drinking water extracts reveal very clean NPD chromatograms only showing very few peaks.

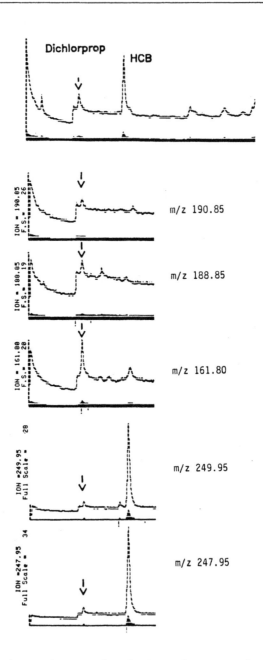

Fig. III.1.–13 Mass fragmentograms of a wheat sample, measured at five specific ions for dichlorprop [20]

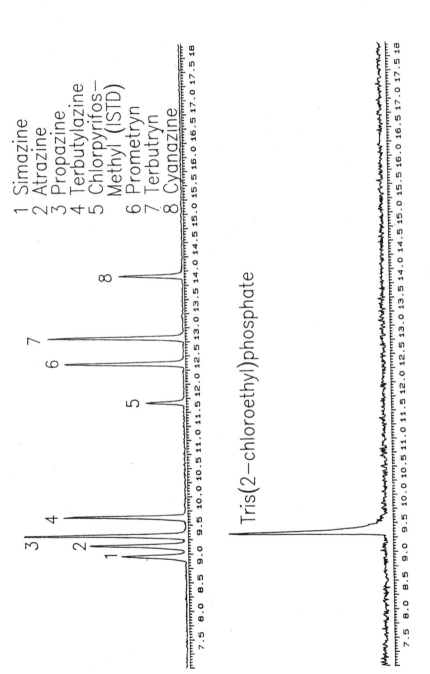

1 Simazine
2 Atrazine
3 Propazine
4 Terbutylazine
5 Chlorpyrifos—
  Methyl (ISTD)
6 Prometryn
7 Terbutryn
8 Cyanazine

Tris(2−chloroethyl)phosphate

Fig. III.1.–14 Gas chromatographic separation of triazine herbicides and tris(2-chloroethyl)phosphate on a 30 m DB-5 fused silica column [57]

Unfortunately, another compound, giving a strong signal on a nitrogen-phosphorus detector, tris(2-chloroethyl-)phosphate almost co-elutes with atrazine and simazine. Figure III.1.–14 shows the elution order of the most important triazine herbicides and the co-eluting interfering compound on a 30 m DB-5 capillary column [57].

In the past, tris(2-chloroethyl-)phosphate, widely used as a flame retardant in paints was occasionally determined in well water [57]. This compound is probably released from the interior of the well shaft or the pump itself. Regardless of its source, this compound may give the false pretention of triazine herbicides.

This example demonstrates the importance of confirming analytical results either on a second capillary column of different polarity or on an independent detection system. GC-MS analysis in SIM mode, for example, offers the feasibility not only to differentiate between the triazine herbicides and the flame retardant in question, but also to quantitate these compounds, even if they are present simultaneously and co-elute completely.

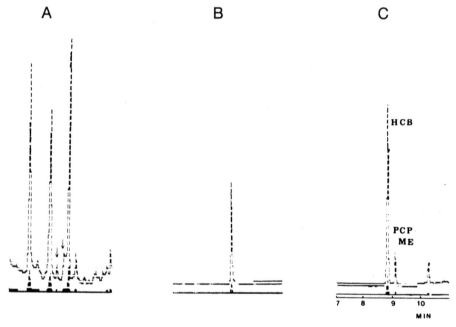

Fig. III.1.–15 Chromatograms of a mushroom extract separated on a 30 m SE-30 glass
capillary column [19]
    A = Total ion current chromatogram
    B = Mass fragmentogram for ion m/z 265 (PCP-Me)
    C = Mass fragmentogram for ion m/z 284 (HCB+PCP–Me)

5.3 Fungicides

Pentachlorophenol (PCP) is a pesticide which acts as a herbicide as well as a fungicide. It is used as a contact herbicide to control weeds, especially in rice and soya cultivation.

More important is, however, the wide-spread use of PCP as a wood preservative because of its fungicidal effects. This application can result in residues e.g., if mushrooms are cultivated in wooden boxes, which had been impregnated with pentachlorophenol.

MEEMKEN et al. determined PCP levels up to 0.2 mg/kg (ppm) in fresh mushrooms [19]. The PCP levels in the wooden boxes employed for cultivation even ranged up to 3.9 g/kg.

The analysis was accomplished by combined GC-MS using selected ion monitoring and electron impact ionization. Chromatographic separation was performed after derivatization with diazomethane on a 30m SE-30 glass capillary column. Hexachlorobenzene (HCB) served as internal standard. Typical chromatograms are depicted in Figure III.1.−15. The topmost chromatogram shows the total ion current of the mass range m/z 33−350. The retention times of PCP-methyl ether and HCB are indicated by arrows. Application of selected ion monitoring to the same extract resulted in traces B and C, which represent the mass fragmentograms for the ions m/z 265 and 284, respectively, typical of PCP-methyl ether and HCB.

The chromatograms being almost free of interfering signals, demonstrate the powerful feasibilities of mass selective detection in residue analysis. Confirmation of the result was performed by measuring four ions, which revealed the same relative intensities as the corresponding reference compound.

It should be noted that the yield in the mushroom production increased significantly after the producer was urged to reduce the pentachlorophenol level in the wooden boxes [19].

6 Summary

Residue analysis has made great progress in the past few years. The introduction of inert glass and fused silica capillary columns has increased chromatographic resolution dramatically. Multidetector retention index standards now offer the possibility of creating large index libraries for almost all pesticides on various stationary phases and different detectors, a valuable help for regulatory pesticide analysis.

Using on-column or programmed temperature vaporizer (PTV) injection techniques, thermally labile substances, such as aldicarb, which are subject to decomposition in hot glass inserts of conventional injection ports, can now be analyzed with high sensitivity.

Although the performance of detectors has improved, a certain trend to mass selective detection can be observed because application of this technique not only reveals a retention time, but also gives important structural information of the compound. Moreover, time- and labor consuming clean-up procedures can often be circumvented due to the high selectivity of mass selective detectors.

Whereas in the past, emphasis was mainly given to the automation of the analytical measurement, nowadays great attempts are made to automate the extraction of samples and the purification of extracts. As a consequence the first reports on the successful use of robots for this purpose have appeared.

All these sophisticated improvements have provided the analyst with powerful tools, enabling him to face the various problems related to pesticide analysis.

Acknowledgement: The author is grateful to Dr. J. J. RYAN, Health and Welfare Canada, for many valuable hints and for critically reading the manuscript.

References

1. THIER, H.-P., FREHSE, H. (1986) "Rückstandsanalytik von Pflanzenschutzmitteln", Georg Thieme Verlag, Stuttgart–New York

2. PETZ, M. (1984) Z. Lebensm. Unters. Forsch. **180**: 267–279

3. ZIEGLER, W., HAHN, M., WALLNOFER, P. R. (1986) Dtsch. Lebensm. Rundsch. **82**: 361–364

4. LANGEBARTELS, C., HARMS, H. (1985) Ecotoxicology and Environmental Safety **10**: 268–279

5. HUTBER, G. N., LORD, E. I., LOUGHMAN, B. C. (1978) J. of Experimental Botany **29**: 619–629

6. FÜRST, P. (1982) PhD Thesis, University of Münster, West Germany

7. SPECHT, W., TILLKES, M. (1980) Fresenius Z. Anal. Chem. **301**: 300–307

8. SPECHT, W., TILLKES, M. (1985) Fresenius Z. Anal. Chem. **322**: 443–455

9. STALLING, D. L., TINDLE, R. C., JOHNSON, D. L. (1972) J. Assoc. Off. Anal. Chem. **55**: 32–38

10. MEEMKEN, H.-A., HABERSAAT, K., GROEBEL, W. (1977) Landwirtsch. Forsch. Sonderheft **34/1**: 262–272

11. FÜRST, P., MEEMKEN, H.-A., GROEBEL, W. (1986) Chemosphere **15**: 1977–1980

12. NORSTRÖM, R. J., SIMON, M., MULVIHILL, M. J. (1986) Int. J. Environ. Anal. Chem. **23**: 267

13. STALLING, D. L., SMITH, L. M., PETTY, J. D., HOGAN, J. W., JOHNSON, J. L., RAPPE, C., BUSER, H.-R. (1983) in "Human and Environmental Risks of Chlorinated Dioxins and Related Compounds" (Tucker, R. E., Young, A. L., Gray, A. P., Eds.), Plenum Publishing Corporation, p 221–240

14. PETZ, M., MEETSCHEN, U. (1987) Z. Lebensm. Unters. Forsch. **184**: 85

15. HOLTMANNSPÖTTER, H., THIER, H.-P. (1982) Dtsch. Lebensm. Rundsch. **78**: 347

16. STORHERR, R. W., WATTS, R. R. (1965) J. Assoc. Off. Anal. Chem. **48**: 1154

17. PFLUGMACHER, J., EBING, W. (1979) Landwirtsch. Forsch. **32**: 82–87

18. EICHNER, M. (1978) Z. Lebensm. Unters. Forsch. **167**: 245–249

19. MEEMKEN, H.-A., FÜRST, P., HABERSAAT, K. (1982) Dtsch. Lebensm. Rundsch. **78**: 282–287

20. MEEMKEN, H.-A., RUDOLPH, P., FÜRST, P., (1987) Dtsch. Lebensm. Rundsch. **83**: 239–245

21. MURPHY, P. G. (1972) J. Assoc. Off. Anal. Chem. **55**: 1360

22. SMREK, A. L., NEEDHAM, L. L. (1982) Bull. Environm. Contam.Toxicol. **28**: 718–722

23. MES, J. (1988) Personal communication

24. DICKES, G. J., NICHOLAS, P. V. (1976) "Gas Chromatography in Food Analysis", Butterworth & Co., London

25. ZWEIG, G., SHERMA, J. (Eds) (1984) "Handbook of Chromatography: Pesticides and Related Organic Chemicals", Vol. I, CRC Press, Boca Raton

26. DRESSLER, M. (1986) "Selective Gas Chromatographic Detectors", Elsevier Science Publishers, Amsterdam

27. VIETHS, S., BLAAS, W., FISCHER, M., KRAUSE, C., MATISSEK, R., MEHLITZ, I., WEBER, R. (1988) Z. Lebensm. Unters. Forsch. **186**: 393

28. FÜRST, P., KRÜGER, C., HABERSAAT, K., GROEBEL, W. (1987) Z. Lebensm. Unters. Forsch. **185**: 17

29. DELEU, R., COPIN, A. (1984) J. High Res. Chromatogr. **7**: 338

30. STAN, H.-J., GOEBEL, H. (1983) J. Chromatogr. **268**: 55

31. STAN, H.-J., MROWETZ, D. (1983) J. Chromatogr. **279**: 173

32. KÜDERLI, F. K. (1976) Chemische Rundschau **29**: 1–5

33. SCHUBERT, R. (1985) GIT Fachz. Lab. **29**: 1175–1177

34. HARRISON, A. G. (1983) "Chemical Ionization Mass Spectrometry", CRC Press, Boca Raton

35. WHITE-STEVENS, R. (ed) (1971) "Pesticides in the Environment", Marcel Dekker, New York

36. MELNIKOV, N. N. (1971) "Chemistry of Pesticides", Springer Verlag, New York–Heidelberg–Berlin

37. The FDA Surveillance Index (with supplements) (1981) US Food and Drug Administration, Washington

38. Pesticide Residues in Food, FAO Plant Production and Protection Papers, FAO, Via delle Terme di Caracalla, 00100 Rome

39. Reviews of Environmental Contamination and Toxicology, Springer Verlag, New York–Heidelberg

40. FÜRST, C. (1988) Unpublished data

41. ZOTOV, L. N., GOLOVKIN, G. V., GOLOVNYA, R. V. (1981) J. High Res. Chromatogr. **4**: 6

42. HALL, G. L., WHITEHEAD, W. E., MOURER, C. R., SHIBAMOTO, T. (1986) J. High Res. Chromatogr. **9**: 266

43. MANNINEN, A., KUITUNEN, M. L., JULIN, L. (1987) J. Chromatogr. **394**: 465

44. HALANG, A., LANGLAIS, R., KUGLER, E. (1978) Anal. Chem. **50**: 1829

45. Nordion Analytical News (1988) Nordion Instruments Oy LTD., Helsinki

46. STAN, H.-J., MROWETZ, D. (1983) J. High Res. Chromatogr. **6**: 255

47. BERTSCH, W. (1978) J. High Res. Chromatogr. **1**: 85

48. BERTSCH, W. (1978) J. High Res. Chromatogr. **1**: 187

49. BERTSCH, W. (1978) J. High Res. Chromatogr. **1**: 289

50. STAN, H.-J. (1988) Lebensm. Chem. Gerichtl. Chem. **42**: 31

51. DEANS, D. R. (1968) Chromatographia **1**: 18

52. MÜLLER, F., ORLEANS, WEEKE, F. (1982) LaborPraxis **6**: 462

53. FÜRST, P. (1986) Unpublished data

54. KWOK, K. S., VENKATARAGHAVAN, R., MC LAFFERTY, F. W. (1973) J. Am. Chem. Soc. **95**: 4185

55. HARAKI, K. S., VENKATARAGHAVAN, R., MC LAFFERTY, F. W. (1981) Anal. Chem. **53**: 386

56. COCHRANE, W. P. (1979) J. Chrom. Science **17**: 124

57. FÜRST, P. (1988) Unpublished data

III.2 Contaminant Analysis

P. Fürst

1 Introduction

In the past few years contaminant analysis has become of increased public interest. This is due to the fact that, in addition to pesticides as "classical residues", environmental pollutants like polychlorinated biphenyls (PCB), polybrominated biphenyls (PBB), polychlorinated dibenzo-p-dioxins (PCDD) and dibenzofurans (PCDF) have become issues of major concern. Moreover, using the latest technology modern analytical instrumentation nowadays offers the feasibility to penetrate into concentration levels of ng/kg (ppt) and even pg/kg (ppq). These extreme detection limits now enable the identification of compounds which have contaminated the environment for years but which could not be detected formerly for the lack of appropriate analytical equipment. This is clearly demonstrated by the present day determination of polychlorinated dibenzo-p-dioxins and dibenzofurans in human milk and foodstuffs.

On the other hand, measurements at these extreme low levels demand a high degree of critical faculty from the analyst with respect to the results and the consequences from his investigations.

Dependent on the type of foodstuff, whether it be of vegetable or animal origin, the analyst dealing with food control has to consider a variety of contaminants. Table III.2.–1, while not claiming to be complete, shows some of the most common classes of compounds possibly leading to a contamination of foodstuffs.

All these compounds, whether used for special industrial purposes or formed as unwanted by-products during synthesis of other technical substances, have been identified in the environment. Hence one may expect that these classes, if they have not already, will also find their way into the food chain and, because of their lipophilic nature, probably lead to a considerable bioaccumulation.

2 Extraction and Clean-up

For extraction and clean-up of the samples basically the same procedures are used, which have already been described in the chapter on pesticide analysis (III. 1).

Due to the lipophilic nature of most contaminants resulting in storage and accumulation in fatty tissues, the first analytical step normally consists in extraction of the lipids along with the substances of interest. This extraction also contains the persistent organochlorine pesticides, which are normally present in a large surplus.

Table III.2.–1 Common environmental pollutants which can contaminate food samples

Polychlorinated Biphenyls

Polychlorinated Biphenylethers

PCB – Substitutes

Polychlorinated Biphenylenes

Polybrominated Biphenyls

Polybrominated Biphenylethers

Polychlorinated Dibenzo-p-dioxins

Polychlorinated Dibenzofurans

Polybrominated Dibenzo-p-dioxins

Polybrominated Dibenzofurans

Mixed-halogenated Dibenzo-p-dioxins

Mixed-halogenated Dibenzofurans

Polychlorinated Naphthalenes

Polybrominated Naphthalenes

Polychlorinated Terphenyls

Polyhalogenated Xanthenes

Polyaromatic Hydrocarbons

Polyhalogenated Hydrocarbons

Polychlorinated Fluorenes

Polyaromatic Chloroethers

Excellent fat solubility and similar physico-chemical properties of many contaminants make analysis more difficult because partitioning into separate fractions often cannot be achieved. Obviously this makes great demands on the analytical determination.

Removal of lipids is performed mostly by using gel permeation chromatography (GPC), a mild and universal clean-up procedure [1] or by degradation with concentrated sulfuric acid. The latter procedure does not only remove lipids but also other oxidizable components. This may be achieved by partitioning an extract between sulfuric acid and an organic solvent, like hexane [2] or by passing the extract, dissolved in a suitable solvent, through a column containing sulfuric acid coated silica gel [3]. This procedure is highly effective and therefore has found widespread use, especially for the determination of polychlorinated biphenyls (PCBs) and polychlorinated dibenzo-p-dioxins (PCDDs) and dibenzofurans (PCDFs).

Depending on the compounds to be analyzed and their expected concentration in the sample, further clean-up steps may differ considerably. For example, if food samples have to be analyzed for residues of PCBs, normally further clean-up simply consists of separation from organochlorine pesticides, which may be present at a similar level. However, analysis for residues of PCDDs and PCDFs, which are present in food samples at concentrations three to six orders of magnitude lower than PCBs, actually requires meticulous clean-up steps in order to remove the bulk of co-extractives, which might probably affect the analytical measurement.

Besides the commonly used adsorbants, Florisil, alumina and silica gel, special carbon types have become more important in recent years. Although various clean-up procedures employing carbon have been described in the past [4, 5], this adsorbant gained special importance only due to a particular pretreatment. Some carbon types effect the separation of planar and co-planar substances from non-planar ones. This is of special interest for the analysis of the planar PCDDs and PCDFs for which separations from non-planar lipophilic pollutants, like most polychlorinated biphenyls on Amoco PX-21 (Amoco Research Corp. Chicago, Ill.) Degusorb BK (Degussa) and Carbopack C (Supelco) have been described [6–10].

The principle of carbon column enrichment consists of a strong retention of the planar and co-planar compounds, which can only be removed by elution with planar solvents, like toluene or benzene. Because of their small particle size, easily leading to blockage of the column, it is necessary to disperse the carbon into shredded glass fiber filters [6] or to mix it with support materials like Celite 545 [10].

The unique properties of these carbon types will presumably lead to a widespread use in residue analysis in the future. Recently, new applications have been reported for the separation of co-planar polychlorinated [11] and polybrominated biphenyls [12] from non-planar ones as well as for the determination of benzo(a)pyrene in smoked foodstuffs [13].

In order to improve analytical precision it has proven good to fortify the samples prior to extraction with ^{13}C-labelled or deuterated compounds. These surrogates are considered as ideal internal standards because they behave similar to the corresponding native compounds during the analysis.

Although gas chromatographic separation of labelled and deuterated compounds from their corresponding native analogues is not possible, differentiation can easily be achieved using combined gas chromatography-mass spectrometry (GC-MS) with the aid of selected ion monitoring (SIM), based on the different mass fragments. For this, one gets an informative idea on the recoveries of the analytes in question.

3 Detection

Because most contaminants are polyhalogenated compounds their determination can be achieved with high sensitivity by electron capture detection (ECD). For the analysis of PCBs in food samples, the ECD is the detector or choice. However, ECD analysis may become problematic if PCBs and PCB-substitutes are present simultaneously or the concentration of contaminants is very low. In this case GC-MS is employed preferentially. Application of the latter technique especially in combination with selected ion monitoring (SIM), not only reveals a retention time but also gives important information on the presence of specific mass fragments, which improves the selectivity of the analytical result. Moreover, modern GC-MS systems nowadays offer the feasibility of measuring compounds at levels of 100 femtogram and lower. Details of GC-MS measurements have already been described in the chapter on pesticides (III. 1).

A variety of different mass spectrometrical techniques have been used to perform analyses of environmental contaminants. The most common types are quadrupole mass spectrometers (LRMS) and double-focussing mass spectrometers (HRMS). Whereas LRMS systems are operated at unit mass resolution < 1 000 (low resolution, LR), the latter ones are often applied at a mass resolution of 10 000 (high resolution, HR).

Sensitivity and selectivity of LRMS can dramatically be increased by employing negative chemical ionization (NCI).

In the past few years occasionally the question has been raised whether high resolution mass spectrometry is mandatory for environmental analysis, especially for the determination of polychlorinated dibenzo-p-dioxins and dibenzofurans. Actually, a round robin study organized by the World Health Organization (WHO) showed, at least for the analysis of PCDDs and PCDFs in human milk, that the applied clean-up procedures and the experience of the analyst are more important than the MS-equipment available [14]. This finding was also substantiated by LINDSTRÖM [15] who demonstrated that PCDD and PCDF analysis in milk samples using low resolution mass spectrometry after high performance clean-up revealed the same accuracy as medium to high resolution mass spectrometry.

However, for the future it can not be excluded that use of high resolution mass spectrometry will become absolutely necessary if investigations have to be performed either at levels far below the present ones or in the simultaneous presence of compounds generating almost the same mass fragments.

More recently two new MS techniques have been reported: MS/MS and Ion Trap Detection (ITD).

Use of MS/MS systems may probably allow interference-free determination of contaminants in complex samples with reduced clean-up. A round robin study involving GC-MS/MS, GC-LRMS and GC-HRMS was designed to determine the feasibility of using GC-MS/

MS techniques for the analysis of PCDDs and PCDFs in difficult environmental samples [16]. Preliminary results showed however, that even this highly effective method requires a certain degree of sample clean-up.

Although the principle of ion trapping has been known since a long time, ion trap detectors became commercially available only a few years ago [17]. The latest ITD version equipped with the enhanced automatic gain control software provides a considerable improvement in sensitivity. As a result of this new acquisition technique, the sensitivity attainable in the linear scanning mode is almost comparable to that in selected ion monitoring (SIM), which is normally employed in trace analysis. Hence the full mass spectral information important for structure elucidation of unknown compounds can be combined with the high sensitivity of SIM.

4   Analytical Determination of Various Compound Classes

4.1   Polychlorinated Biphenyls

Polychlorinated biphenyls (PCBs) comprise 209 chlorinated congeners of biphenyl with one to ten chlorine atoms (Fig. III.2.–1).

$$x + y = 1 - 10$$

PCB

Fig. III.2.–1 Structure of polychlorinated biphenyls (PCBs)

PCBs are the major components in products marketed as Arochlor 1221, 1242, 1254, 1260, Kanechlor 300, 400, 500, 600 and Clophen A30, A40, A50 and A60.

Owing to their excellent physical properties, polychlorinated biphenyls have enjoyed widespread use as dielectric fluids for capacitors and transformers because they possess excellent thermal conductivity while their electric conductivity is extremely low [18]. They have also played a significant role as hydraulic fluids in underground mining, solvent extenders, flame retardants and organic diluents [19]. The widespread use of technical PCB mixtures coupled with improper disposal practices has meanwhile led to their global distribution. Moreover, disposal of PCB containing wastes may represent a severe problem because incomplete incineration results in formation and emission of PCDDs and

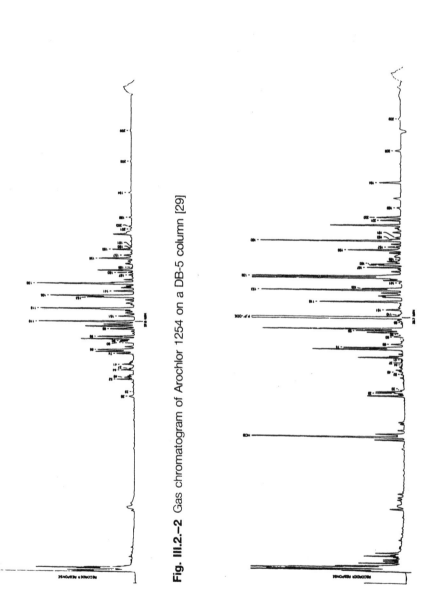

Fig. III.2.–2 Gas chromatogram of Arochlor 1254 on a DB-5 column [29]

Fig. III.2.–3 Gas chromatogram of a composite human milk sample from Canada on a DB-5 column [29]

PCDFs. These toxic contaminants have also been introduced into the environment from fires of capacitors and transformers [20, 21].

Due to their lipophilic nature, PCBs are stored in fatty tissues and bioaccumulated in the food chain.

Until recently, most investigations concerning PCB analysis have been carried out by using packed columns. The concentrations were then estimated by comparing specific signals, which were not superimposed with known organochlorine pesticides with corresponding peaks from a technical PCB mixture. Actually this calculation does not take into account that a lot of PCBs are metabolized in biota while others are more resistant to biodegradation and therefore accumulate.

On the other hand, it is difficult to compare results from different countries due to the fact that various technical mixtures, which serve as reference standards, differ to a certain degree. Hence PCB data from biological samples originating from investigations using packed columns can only be considered as semiquantitative estimates of the total PCB level [22].

Moreover, biologic and toxic effects of PCBs are highly dependent on structure [23, 24]. The most toxic PCBs known are 3,4,3',4'-tetra-, 3,4,5,3',4'-penta- and 3,4,5,3',4',5'-hexachlorobiphenyl [25, 26]. Toxicity of these congeners is caused by their co-planar structure making them approximately isostereomer with 2, 3, 7, 8-tetrachlorodibenzo-p-dioxin (2, 3, 7, 8-T_4CDD). Addition of a single ortho-chloro substituent to the co-planar PCBs diminishes their toxic potency [22].

From what was mentioned above it follows that the use of capillary gas chromatography for isomer-specific analysis of PCBs in biological samples is mandatory in order to compare analytical results and to make a meaningful risk assessment.

In addition, some countries plan to issue PCB regulations based on single congeners. For example, in the Federal Republic of Germany maximum residue limits for PCBs with numbers 28, 52, 101, 138, 153 and 180 have been passed [27]. The numbers refer to the BALLSCHMITER and ZELL numbering system, which has been internationally accepted [28].

Figure III.2.–2 shows the gas chromatographic analysis of the technical PCB mixture Arochlor 1254 on a DB-5 column [29]. A typical chromatogram of a composite human milk extract originating from Canada is depicted in Figure III.2.–3 [29]. These two chromatograms clearly demonstrate the striking differences in the congener pattern for a technical mixture and an extract of a biological specimen which has undergone metabolism.

In human milk samples, congener numbers 74, 118, 138, 153 and 180 make up approximately 67−75 % of total PCBs with numbers 153 and 138 as predominant components [30−32].

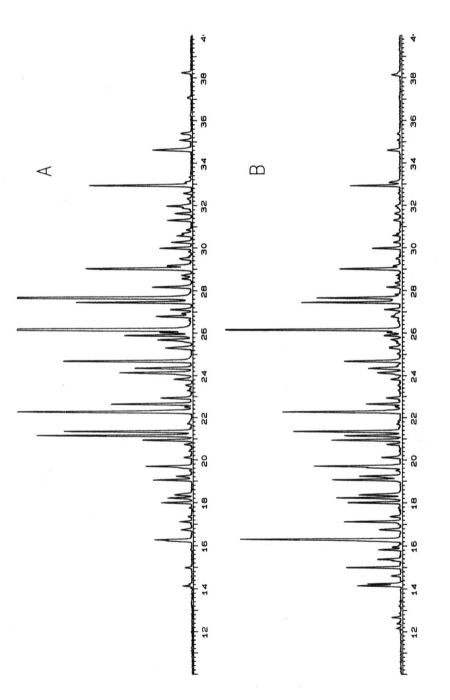

Fig. III.2.–4 Gas chromatograms of Clophen A40 **(A)** and a fish specimen **(B)** caught in an area with extensive mining [33]

Some fish samples, however, may reveal chromatograms, which resemble closely technical mixtures. Figure III.2.–4 shows chromatograms of the technical mixture Clophen A40, a common PCB product in the Federal Republic of Germany and a fish extract, both separated on a DB-5 column under the same conditions [33]. The chromatograms exhibit great similarities with respect to the peak pattern. So one can assume that a considerable biodegradation in the fish has not occurred so far. This might be due to the fact that the fish was caught in an area of extensive mining where PCB-mixtures are widely used as hydraulic fluids in underground mining. Although this use is listed as a closed application, the hydraulic fluid seems to be discharged. Once set free underground, it can enter the environment through pit waters, mine outputs and ventilation systems [18].

The ECD response is not only dependent on the type and number of halogen atoms but also on their position in the molecule. This may lead to difficulties when quantifying complex PCB-mixtures because different isomers of a given degree of chlorination cannot be calculated using the same response factor. Consequently a precise quantification of PCBs is only possible if the compound of interest is available as a reference standard and a calibration can be performed. Obviously, this causes some problems because only a limited number of PCB-congeners are commercially available.

The problem can be circumvented by using a flame ionization detector (FID) for an "intermediate" quantification of a technical mixture. This is feasible because the FID response is a function of the number of chlorine atoms in the molecule. With the aid of a calibration standard containing one isomer per degree of chlorination, response factors can be calculated for each isomer group. Subsequently these factors can be used for quantification of all congeners in a technical mixture [34, 35].

This approach basically requires knowledge of the number of chlorine atoms of each component, which can be performed by GC-MS. Finally the data gained from the FID measurements can be used to calculate the response factors on the ECD.

Recently MULLIN et al. [36] synthesized all 209 different PCB congeners and reported their relative response factors and relative retention times in relation to octachloronaphthalene. The measurements were performed on a capillary column coated with SE-54. These data may serve as a valuable basis for an isomer-specific PCB analysis for those laboratories which only have access to a very limited number of standard compounds.

As mentioned earlier, the most toxic PCB congeners are non-orthochlorine substituted and therefore can adopt a co-planar conformation. Their co-planar structure, allows the separation from the bulk of other PCBs by means of a carbon column [11]. Some carbon columns offer the feasibility to strongly retain planar and co-planar compounds while non-planar substances pass. Finally the co-planar PCBs can be removed from the column by elution with benzene or toluene resulting in a high enrichment of the toxic congeners.

Up to the present there are only very few data on levels of coplanar PCBs in the environment [11] but these results indicate that we have to contend with a widespread distribution of these highly toxic environmental pollutants.

4.2 PCB – Substitutes

In the last few years the knowledge of the potential health risk of PCBs has led to efforts to replace them. Meanwhile suitable substitutes are available for almost all purposes [18]. One group of substances that has replaced PCBs to a large extent are tetrachlorobenzyltoluenes (TCBTs). The structure of these compounds is shown in Figure III.2.–5.

$$CH_3$$

$$-CH_2-$$

$$Cl_x \quad x + y = 4 \quad Cl_y$$

Fig. III.2.–5 Structure of tetrachlorobenzyltoluenes (TCBTs)

TCBTs are the major components in products marketed as Ugilec 141, which is a mixture of pure TCBTs, and Ugilec T, additionally containing about 40 % of trichlorobenzenes.

Because of their good technical properties, comparable to those of PCBs, and in accordance with safety requirements, TCBT containing products have found widespread use as hydraulic fluids in underground mining.

TCBTs released underground can enter the environment via pit waters, mine outputs or ventilation systems and as a consequence they can contaminate the food chain. Only recently these compounds have been determined in fish from areas of extensive mining [37–39].

Analysis of TCBTs in food samples is complicated by the simultaneous presence of PCBs. Due to their similar polarity, it is almost impossible to separate these two classes selectively. Moreover, they possess comparable volatility, which leads to co-elution in gas chromatographic analysis.

Figure III.2.–6 compares chromatograms of the technical mixture Ugilec 141 with parts of the PCB mixture Clophen A30/A60 and some selected PCB congeners, separated on a 30 m DB-5 fused silica column [40]. These chromatograms clearly demonstrate that TCBTs elute from the column between the PCB congeners with Ballschmiter numbers 101 and 138. For this, GC-ECD analysis of tetrachlorobenzyltoluenes is adversely affected when PCBs are present simultaneously.

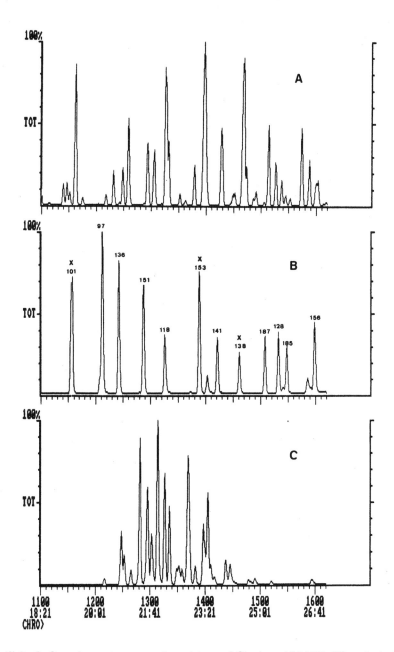

Fig. III.2.–6 Gas chromatograms of a mixture of Clophen A30/A60 **(A)**, selected PCB congeners **(B)** and Ugilec 141 **(C)** separated on a DB-5 column [40]

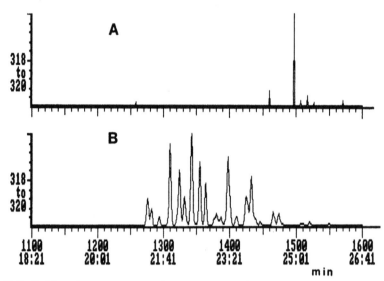

Fig. III.2.–7 Mass fragmentograms (m/z 318–320) of a mixture of Clophen A30/A60 **(A)** and Ugilec 141 **(B)** on a DB-5 column [37].

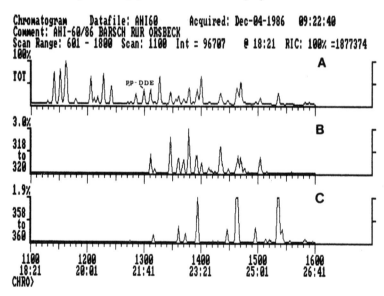

Fig. III.2.–8 Chromatograms of a pike extract [40]

    (A) Total ion chromatogram

    (B) Mass fragmentogram m/z 318–320 (TCBTs)

    (C) Mass fragmentogram m/z 358–360 (Hexachlorobiphenyls)

On the other hand, isomer-specific PCB determination with the aid of GC-ECD may reveal false positive results due to residues of TCBTs. This may become problematic in so far as the German Government recently issued maximum residue limits for six PCB congeners including numbers 101, 138 and 153 [27].

Because both groups generate different mass fragments, mass selective detection permits an accurate determination of both TCBTs and PCBs, even in the presence of a large excess of the other components.

Figure III.2.–7 shows mass fragmentograms for the ion range m/z 318–320 of a mixture of Clophen A30/A60 and Ugilec 141 obtained with an ion trap detector. It can be clearly seen that the specific fragments chosen for the determination of TCBTs do not interfere with PCBs, even if the PCB concentration is 100 times higher. The only interfering signal so far recognized is p,p,-DDE, which is also detectable within this mass range but elutes directly before the first TCBT isomer.

The simultaneous determination of TCBTs and hexachlorobiphenyls in a pike extract is depicted in Figure III.2.–8. The upper chromatogram represents the total ion current, and the two lower ion traces show the presence of TCBTs (B) and hexachlorobiphenyls (C) reconstructed by choice of characteristic masses for the compounds of interest [40].

Investigations of fish samples from areas with extensive mining revealed TCBT levels up to 25 mg/kg (ppm) calculated as Ugilec 141 based on the edible portion [37–39]. Due to the lack of specific isomers, quantitation was accomplished by an external standard method using Ugilec 141 for calibration. Hence, these levels only reflect an approximation of the real concentrations.

It is interesting that the isomeric pattern differs considerably depending on the fish species. Figure III.2.–9 compares the TCBT-profiles from three different species [38]. The presence of TCBTs was confirmed by acquisition of complete mass spectra.

The relative ratios of the three signals marked by an asterisk were surprisingly constant and independent from the fishing grounds. If this is caused by metabolism or by the fact that these species represent a placid-, predatory- and mud fish can only be speculated.

Because toxicological data are limited a risk assessment of TCBTs is not possible at the moment. Nevertheless, their identification in the food chain should be used as an opportunity to reconsider their application in order to avoid further introduction into the environment, which might probably lead to an ubiquitious distribution and bioaccumulation similar to the PCBs.

4.3 Polybrominated Biphenyls and Biphenylethers

Polybrominated biphenyls (PBBs) and biphenylethers (PBBEs) represent two classes of snythetic chemicals that comprise 209 brominated congeners of biphenyl and biphenylether, respectively, with one to ten bromine atoms (Fig. III.2.–10).

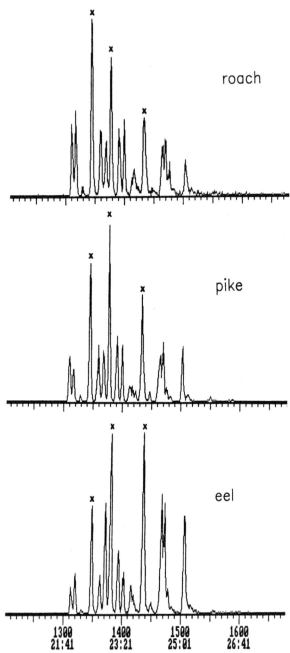

Fig. III.2.–9 TCBT profiles of various fish species [38]

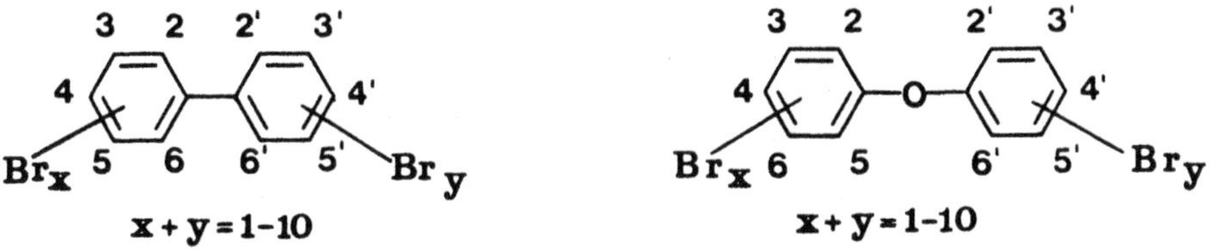

Fig. III.2.–10 Structures of polybrominated biphenyls (PBB) and biphenylethers (PBBE)

PBBs and PBBEs are of great importance as additives to plastics as flame retardants. For this reason these two classes of compounds have found widespread use in various synthetic polymers, particularly in cars, air planes, household appliance and consumer goods. They are also utilized as textile additives. Their addition amounts to approximately 10 % [41].

Production of brominated flame retardants in the USA, Japan and West Europe amounted to 28 000 tons in 1975 [42]. Although actual data are not available, one may estimate an annual increase of 15–17 % [43].

During production and use, PBBs and PBBEs can enter the environment in different ways. While these two compound classes do not have the industrial significance of the PCBs, they are similar with regard to their lipophilic and persistent properties. Hence, after release into the environment, one has to reckon with a bioaccumulation in the food chain.

PBBs became an issue of major concern in 1973 when they were accidentally substituted for magnesium oxide in livestock feed in Michigan leading to an exposure of cattle and people consuming their tainted food products [44]. After the sources had been identified much emphasis was placed on analysis of dairy products and other foodstuffs that were produced in the contaminated area [45, 46]. Moreover, a cohort of about 4 000 potentially exposed farm residents were surveyed in order to investigate any adverse health effects [47].

At that time analyses were performed by gas chromatography with electron capture detection [48]. Quantitation was usually based on the most abundant hexabromo-congener.

However, as with the PCBs, the toxicity of PBBs differs considerably depending not only on the degree of bromination but also on the position of the bromine atoms [49]. Hence, use of capillary columns should be the method of choice in order to effect a congener-specific analysis.

ROBERTSON et al. reported on the synthesis and identification of highly toxic polybromina-ted biphenyls in the fire retardant Firemaster BP-6 [50]. They synthesized 19 PBBs from specific precursors and confirmed the structure by their proton magnetic resonance (HNMR) and mass spectra. Separation of the compounds was carried out on capillary columns coated with SE-54.

Recently KRÜGER reported on the congener-specific determination of PBBs and PBBEs in various foodstuffs and human milk samples from the general population in Germany [12].

The compounds were extracted together with fat and other lipophilic substances either from the freeze-dried material (fish) or, in case of milk, directly by use of organic solvents. Gel permeation chromatography (GPC) on Bio-Beads S-X3 was employed for removal of lipids. To establish, at which times the compounds are eluted from the GPC-column, elution profiles for the active substances as well as for extracts of food samples to be analyzed were determined.

Figure III.2.–11 depicts the elution profile of milk fat and several polybrominated biphenyls on Bio-Beads S-X3 using cyclohexane/ethyl acetate 1/1 as a solvent. It can be clearly seen that all PBBs are well separated from the weighable amount of lipids.

As already mentioned, GPC mainly separates according to the molecular weight. Nor-mally big molecules elute first followed by smaller compounds, which fit better into the gel bed thereby retarding their elution. For this reason, the profile shown in Figure III.2.–11 is somewhat unexpected because the lower brominated biphenyls elute first and deca-bromobiphenyl elutes last. This example points out that the separation on the GPC-column is also influenced by other effects presumably like τ-bond attractions.

After removal of lipids, adsorption chromatography on Florisil deactivated with 1 % water was used to separate PBBs and PBBEs from more polar compounds such as PCDDs and PCDFs. Finally the Florisil-extract was applied to a carbon column in order to frac-tionate into the non-planar and co-planar components. This is of special interest because co-planar PBBs possess a higher toxicity than the other congeners [49].

A DB-5 fused silica column was used for separation of the compounds. Mass spectro-metric detection was performed by use of NCI in the SIM-mode.

The described procedure provides identification and determination of PBBs and PBBEs in the ppt-range even in the presence of an excess amount of PCBs. The application of the developed procedure to the investigation of fish and seal blubber specimens revealed low background levels with typical congener profiles depending on the origin of the sample.

Figure III.2.–12 shows the ion traces for hepta- and hexabromobiphenyls typical of fish specimens caught in the North Rhine-Westphalian river Lippe (West Germany). The numbers above the peaks have been adopted from the Ballschmiter and Zell numbering system for PCBs [28].

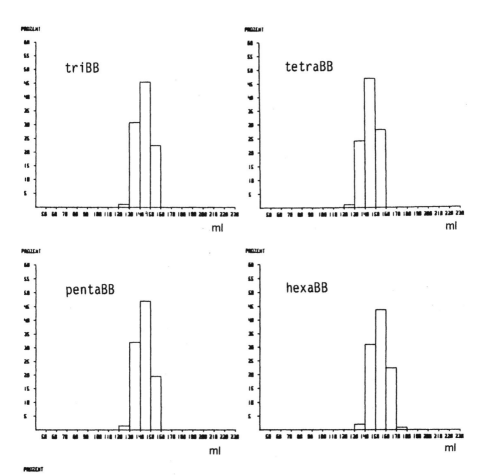

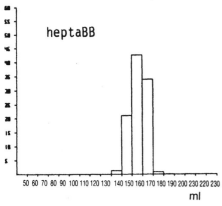

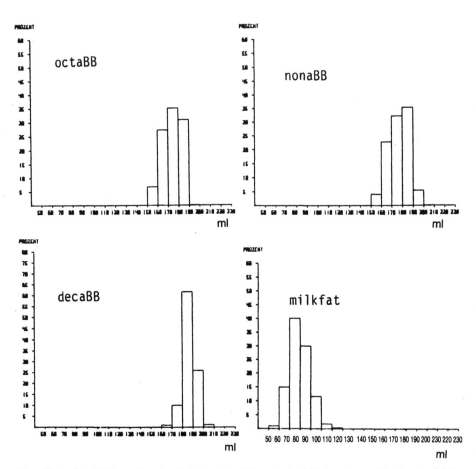

Fig. III.2.–11 GPC profile of milk fat and polybrominated biphenyls on Bio Beads S-X3 with cyclohexane/ethyl acetate 1/1 as elution solvent [12] *(continued)*

Congener depending, the PBB levels ranged between 0.01 and 2 µg/kg (ppb), calculated on a lipid basis.

Whereas the PBBE levels in fish samples from the Baltic Sea and the Atlantic Ocean on average amounted to 25 µg/kg (ppb), their concentration in fish caught in North Rhine-Westphalian rivers ranged up to 500 ppb, each calculated as Bromkal 70–5 and based on lipid.

Besides fish, which are good bioindicators for environmental pollutants, some PBB- and PBBE congeners could also be determined in cow's and human milk. Typical chromato-

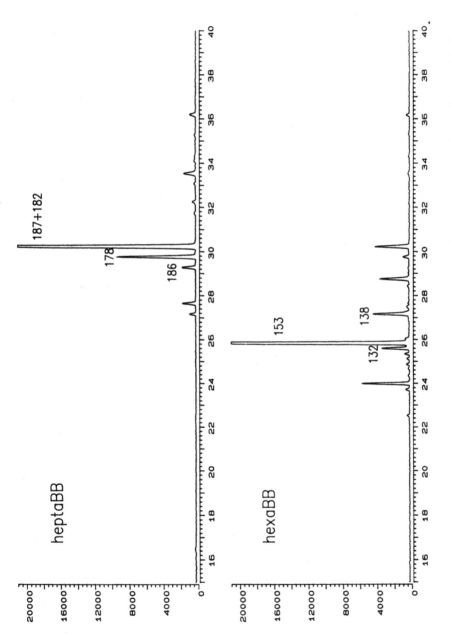

Fig. III.2.–12 Hepta- and hexabromobiphenyls in fish from North Rhine-Westphalian rivers (West Germany) [12]

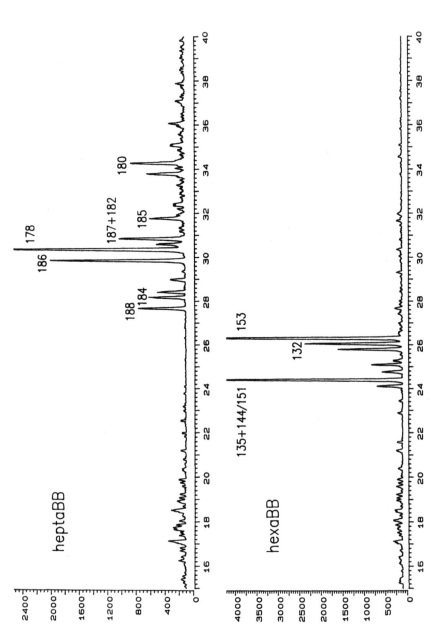

Fig. III.2.–13 Hepta and hexabromobiphenyls in human milk from West Germany [12]

grams of hepta- and hexabromobiphenyls in a human milk extract are depicted in Figure III.2.–13.

The predominant congener identified in cow's and human milk was 2,2',4,4',5,5'-hexabromobiphenyl (PBB # 153) with average levels of 0.03 and 1.0 ppb on a lipid basis, respectively. Higher brominated biphenyls were only found to be peaks of minor concentration.

The predominance of PBB # 153 is not surprising because it is the most abundant congener in the widely used flame retardants Firemaster BP-6 and Firemaster FF-1. In addition, it is known from PCB investigations that halogen substitution in the 2,2',4,4', 5,5'-positions at the biphenyl rings leads to a high bioaccumulation in mammals.

The global distribution of PBBs and PBBEs was demonstrated by analysis of seal blubber from Spitsbergen (Norway) and a breast milk sample from a Chinese woman, which also revealed background levels of these two classes of environmental contaminants [12].

4.4 Polychlorinated Dibenzo-p-dioxins and Dibenzofurans

Polychlorinated dibenzo-p-dioxins (PCDDs) and dibenzofurans (PCDFs) are two series of aromatic ethers with one to eight chlorine atoms per molecule.

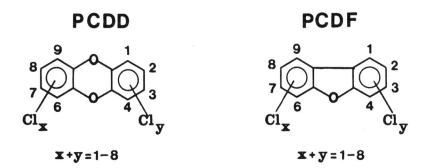

Fig. III.2.–14 Structures of polychlorinated dibenzo-p-dioxins (PCDDs) and dibenzofurans (PCDFs)

Dependent on their degree of chlorination one can distinguish 75 PCDD and 135 PCDF positional isomers, respectively. These two classes of environmental contaminants have been identified as unwanted by-products in various technical products [51–54]. Other major sources comprise municipal and hazard waste incineration [55], paper bleaching [56–58] and exhaust from cars without catalytic converters using leaded gasoline and a dichloroethane scavenger [59]. Moreover, BALLSCHMITER et al. reported on the determination of PCDDs and PCDFs in used motor oils from automobiles [60].

Due to the manifold sources, PCDDs and PCDFs have meanwhile found an ubiquitious distribution. Consequently, they were identified in various biological samples, including adipose tissues, blood, food samples and human milk [8, 61–71]. RAPPE et al. [72] give a comprehensive overview on environmental fate of PCDDs and PCDFs with special regard to sources, levels and isomeric pattern in different matrices.

The toxicity of PCDDs and PCDFs differs considerably. Table III.2.–2 shows the most toxic congeners, often termed the "dirty dozen".

Table III.2.–2 Most Toxic PCDD and PCDF Congeners (Dirty Dozen)

$2,3,7,8 - T_4CDD$	$2,3,7,8 - T_4CDF$
$1,2,3,7,8 - P_5CDD$	$1,2,3,7,8 - P_5 CDF$
$1,2,3,4,7,8 - H_6CDD$	$2,3,4,7,8 - P_5CDF$
$1,2,3,6,7,8 - H_6CDD$	$1,2,3,4,7,8 - H_6CDF$
$1,2,3,7,8,9 - H_6CDD$	$1,2,3,6,7,8 - H_6CDF$
	$1,2,3,7,8,9 - H_6CDF$
	$2,3,4,6,7,8 - H_6CDF$

In general, congeners which are 2,3,7,8-chlorine substituted and possess at least one vicinal proton exhibit a higher toxicity than others. For example, the acute toxicity of 2,3,7,8- and 1,2,3,8-tetrachlorodibenzo-p-dioxin (T_4CDD) differs by a factor of 1 000–10 000 [68]. Hence it follows that an isomer-specific PCDD- and PCDF-determination is mandatory to gain knowledge on the congener profiles as a basis for a meaningful risk assessment.

Although capillary columns provide a high resolution power, it is not possible so far to separate all different PCDD and PCDF congeners using only one column. It rather needs at least two columns of different polarity not only to perform an isomer-specific analysis in incineration samples, like fly-ash, where all congeners with four to eight chlorine atoms are normally present but also for confirmation of results. This is especially important for congeners with 2,3,7,8-chlorine substitution.

Figure III.2.–15 shows the elution order of all 22 tetrachlorodibenzo-p-dioxins on a CP-Sil 88 and an OV-17 column [73]. It can be clearly seen that 2,3,7,8-T_4CDD, the most toxic isomer, is superimposed with 1,2,7,9-T_4CDD on the OV-17 column. For this an unambigious determination of 2,3,7,8-T_4CDD using only this column in incineration samples is not possible. On the other hand cyano-propyl silicone coated columns, like SP 2331 or CP-Sil 88 allow a specific separation of 2,3,7,8-T_4CDD from all other T_4CDD-isomers.

Figure III.2.–16 compares the elution order of pentachlorodibenzofurans (P_5CDF) on a CP-Sil 88 and an OV-17 column [73]. In this case the OV-17 column provides a specific separation of 1,2,3,7,8- and 2,3,4,7,8-P_5CDF, the two P_5CDF isomers with 2,3,7,8-chlorine substitution, from all other isomers, whereas 1,2,3,7,8,-P_5CDF co-elutes with 1,2,3,4,8-P_5CDF on the CP-Sil 88 column.

Recently SWEREV and BALLSCHMITER reported on an isomer-specific separation of 2,3,7,8-chlorine substituted PCDD- and PCDF congeners on a Smectic liquid-crystalline stationary phase [74]. The separation mechanism of this column is dependent on the molecular characteristics of the compound. The length-to-breadth ratio of analytes seems to have a great influence on their retention. For this a unique selectivity for most of the 2,3,7,8-class congeners was observed. One of the most important features is the pronounced retention of 2,3,7,8-T_4CDD, which elutes after all other T_4CDDs. The Smectic column also allows a separation of 2,3,7,8-tetrachlorodibenzofurans (T_4CDF) from the other tetrachloroisomers, which is difficult to achieve with other columns.

Their high selectivity in combination with the good thermostability will probably lead to a wide-spread use of Smectic columns in the future not only for PCDD- and PCDF analyses but also for other planar compounds, such as halogenated naphthalenes and pyrenes.

PCDD- and PCDF analysis became an issue of major concern after these contaminants have been identified in human milk samples from the general population [68]. Table III.2.–3 shows the results of the isomer specific determination of PCDDs and PCDFs in 193 individual human milk specimens from North Rhine-Westphalia (West Germany) [9].

The results document the occurrence of PCDDs and PCDFs in breast milk samples and the presence of a typical pattern. Common characteristic of all specimens analyzed is 2,3,7,8-chlorine substitution. Whereas PCDD normally represents more than 50 % of the total dioxin amount, the levels of the other dioxin congeners decrease with decreasing degree of chlorination.

Compared with PCDDs, the PCDF levels are generally considerably lower. 2,3,4,7,8-P_5CDF usually represents the main compound of the polychlorinated dibenzofurans analyzed so far in human milk.

Meanwhile, it is generally accepted that the major route of exposure to PCDDs and PCDFs is via food. However, the contamination level is highly dependent on the origin of the sample. This is clearly demonstrated by the analysis of fish and aquatic organisms from the Great Lakes in North America where great variations in the PCDD and PCDF levels were observed, indicating the strong influence of point source discharges [75].

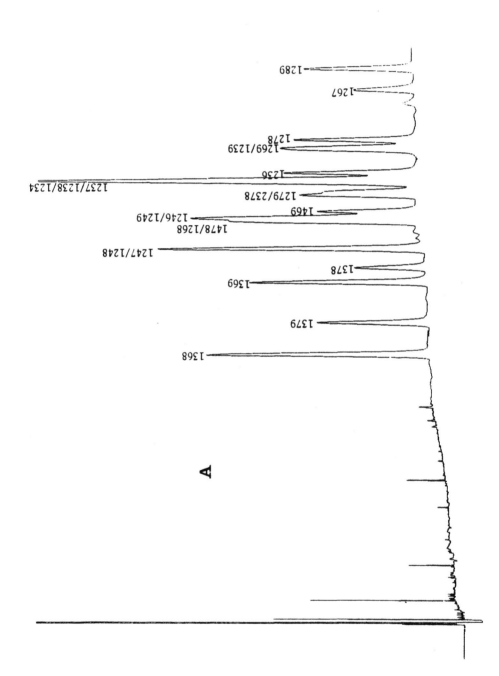

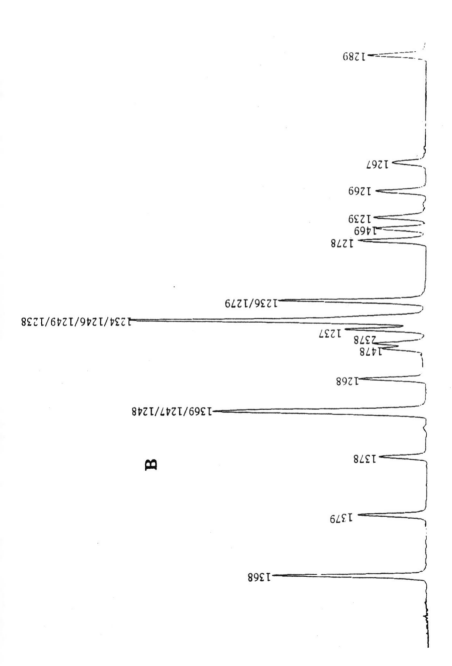

Fig. III.2.–15 Elution order of all tetrachlorodibenzodioxins on an OV-17 **(A)** and CP-Sil 88 column **(B)** [73].

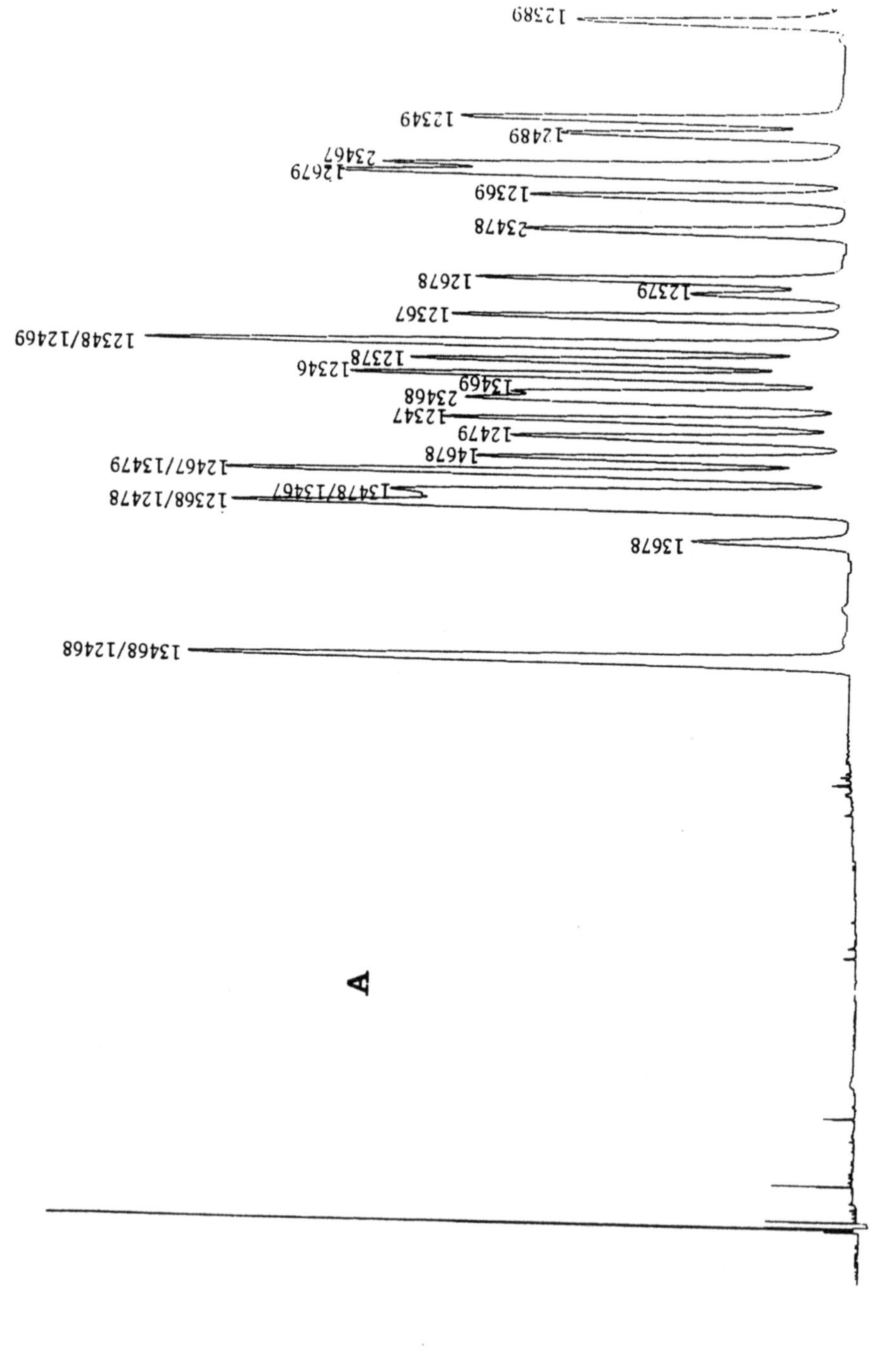
12589
12349
12489
23467
12679
12369
23478
12678
12379
12367
12348/12469
12378
12346
13469
23468
12347
12479
14678
12467/13479
13478/13467
12368/12478
13678
13468/12468
A

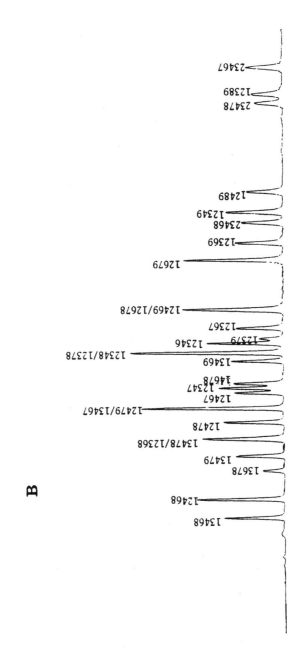

Fig. III.2.–16 Elution order of all 28 pentachlorodibenzofurans from an OV-17 **(A)** and CP-Sil 88 column **(B)** [73]

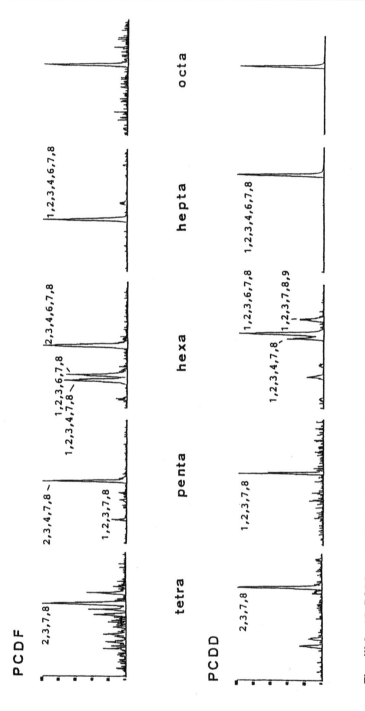

Fig. III.2.–17 PCDD and PCDF profiles of various food samples on a DB-5 column [77]
A Beef

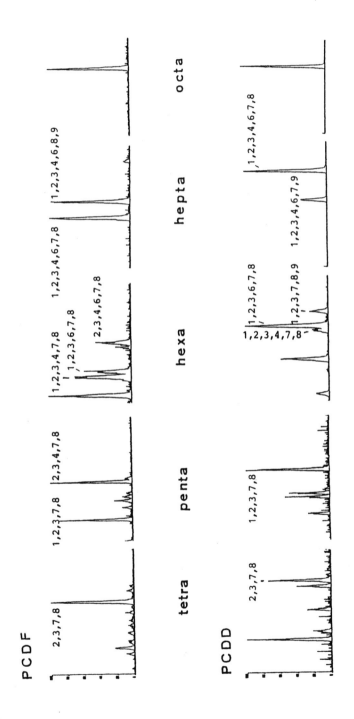

Fig. III.2.–17 PCDD and PCDF profiles of various food samples on a DB-5 column [77].
B Chicken

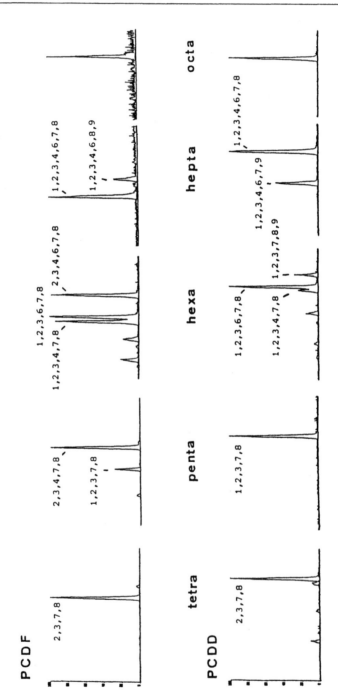

Fig. III.2.–17 PCDD and PCDF profiles of various food samples on a DB-5 column [77]
A Beef, **B** Chicken, **C** Herring

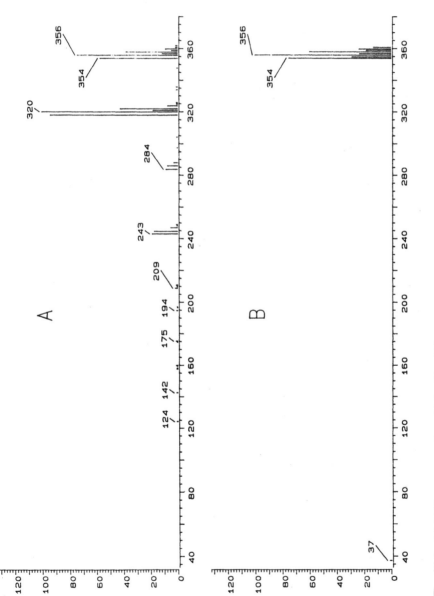

Fig. III.2.–18 NCl mass spectra of tetradifon **(A)** and 1,2,3,7,8-P₅CDD **(B)** [78]

Table III.2.–3 PCDD and PCDF Levels in Human Milk from West Germany [9] (n = 193)

Congener	Amount ng/kg (ppt) calc. on fat basis	
	mean value	range
Octachlorodibenzodioxin	195	13 – 664
1,2,3,4,6,7,8 – H_7CDD	44.7	11 – 174
1,2,3,4,7,8 – H_6CDD	7.6	1 – 24
1,2,3,6,7,8 – H_6CDD	31.2	6 – 123
1,2,3,7,8,9 – H_6CDD	6.6	1 – 21
1,2,3,7,8 – P_5CDD	9.9	1 – 40
2,3,7,8 – T_4CDD	2.9*)	< 1 – 7.9
Octachlorodibenzofuran	8.2	< 1 – 86
1,2,3,4,6,7,8 – H_7CDF	4.8	< 1 – 20
1,2,3,4,7,8 – H_6CDF	7.3	1 – 28
1,2,3,6,7,8 – H_6CDF	5.8	1 – 25
2,3,4,6,7,8 – H_6CDF	3.4	1 – 9
1,2,3,7,8 – P_5CDF	0.7	< 1 – 7
2,3,4,7,8 – P_5CDF	25.1	1 – 67
2,3,7,8 – T_4CDF	2.2	< 1 – 5

*) 82 samples analyzed

RAPPE et al. [72] analyzed crabs and crustacean hepatopancreas from the Swedish west coast and found the highest PCDD and PCDF levels in samples from fiords where effluents from pulp mills using chlorine in their bleaching process are introduced.

The same authors also analyzed cow's milk samples from Switzerland and determined somewhat higher levels in specimens, which were collected in the vicinity of incinerators [76].

Common characteristic of all biological samples analyzed so far is the predominance of 2,3,7,8-chlorine substituted congeners.

Figure III.2.–17 shows the congener profiles of beef, chicken and herring [77]. Only the chromatogram of the chicken extract reveals some signals which do not belong to the toxic 2,3,7,8-class. This may be due to a difference in metabolism in poultry compared to other animals or to a recent exposure to these contaminants.

When analyzing fruits or vegetables for residues of PCDDs and PCDFs, care must be taken because tetradifon, a widely used acaricide, may indicate the presence of penta-chlorodibenzo-p-dioxin [87].

Figure III.2.–18 compares the NCI mass spectra of tetradifon and 1,2,3,7,8-pentachloro-dibenzo-p-dioxin. Although both compounds differ by one chlorine atom, the mass spectra are very similar because the sulfur atom in the acaricide also contributes to the isotopic pattern to a certain degree. Moreover, both compounds almost co-elute on a DB-5 column which may cause problems for SIM measurements. A separation, however, is possible by use of a carbon column, which removes the non-planar acaricide from the planar pentachlorodibenzo-p-dioxins.

This example demonstrates the need for a meticulous clean-up even for those samples which are not expected to be problematic.

Based on an investigation of various food samples and consumption habits, BECK et al. [77] estimated the body burden with PCDDs and PCDFs from food intake in Germany. They calculated a daily intake via food of 24 pg 2,3,7,8-T_4CDD and 93 pg T_4CDD-equivalents according to the equivalent factors proposed by the German Federal Health Office. Whereas fish and fish products amounted to about 30 % of the daily T_4CDD-equivalents intake, vegetables and fruits contributed only 1–5% to the body burden by food intake.

All investigations of biological and food samples revealed a general low background of 2,3,7,8-chlorine substituted PCDDs and PCDFs, which is an indication that these environmental contaminants can no longer be considered as a problem of limited extension but of global concern.

4.5   Miscellaneous Contaminants

Figure III.2.–19 shows some compound classes, which have been released into the environment by manufacturing processes, fires from transformers and capacitors and from municipal and hazard waste incinerators [21, 79–83]. Consequently, one has to reckon with a bioaccumulation because of the lipophilic nature of these environmental compounds.

PAASIVIRTA et al. [83] recently reported on the identification of PCBE, PCPA, PCA, PCV, PCF and PCDHA in salmon and eagle specimens from the Baltic. The levels they measured were in the ppb-range (Fig. III.2.–19).

The analysis is complicated by the fact that some classes generate similar fragments during GC-MS-analysis. For example, trichloropyrenes and tetrachlorodibenzofurans exhibit the same nominal molecular ion (m/z 304), and the molecular ions for penta- and hexachlorobiphenylenes are the same as the nominal masses for $(M+2)^+$ ions of tetra- and pentachlorodibenzo-p-dioxins.

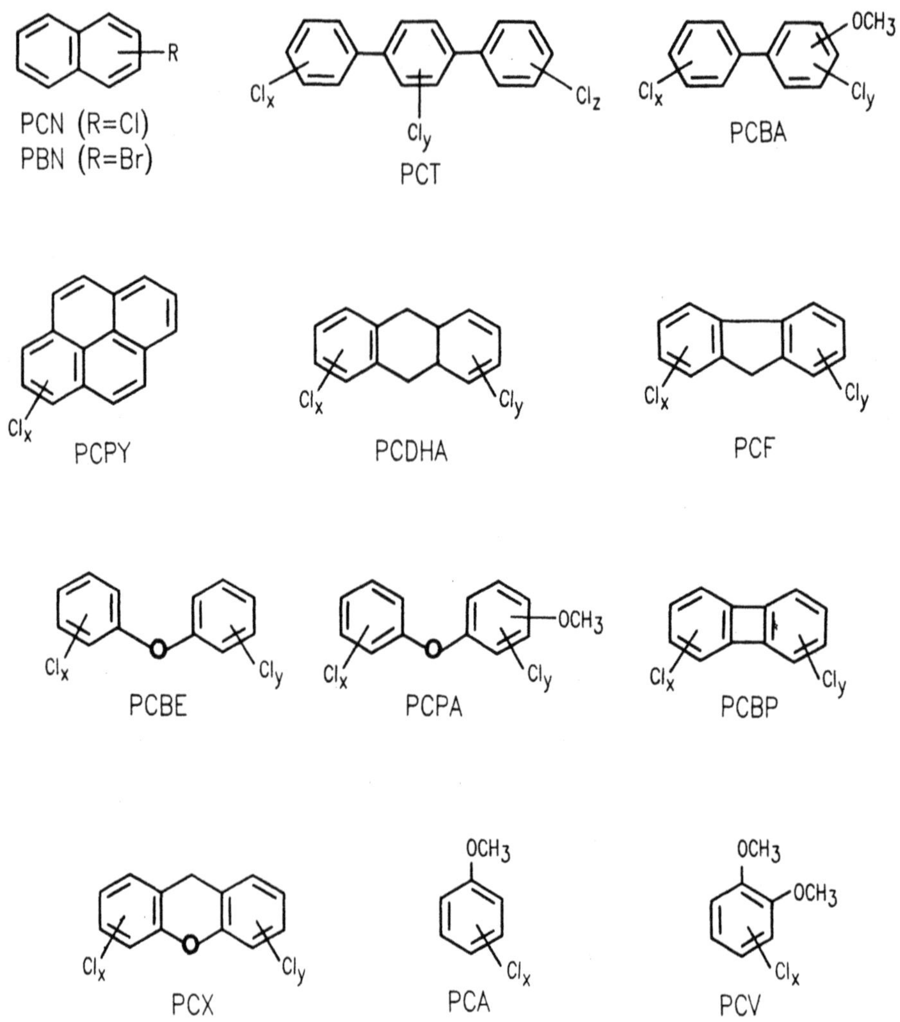

Fig. III.2.–19 Structures of persistent pollutants

PCN	Polychlorinated naphthalenes	PCBE	Polychlorinated biphenylethers
PBN	Polybrominated naphthalenes	PCPA	Polychlorinated phenoxyanisoles
PCT	Polychlorinated terphenyls	PCBA	Polychlorinated biphenylanisoles
PCBP	Polychlorinated biphenylenes	PCX	Polychlorinated xanthenes
PCPY	Polychlorinated pyrenes	PCA	Polychlorinated anisoles
PCDHA	Polychlorinated dihydroanthracenes	PCV	Polychlorinated veratroles
PCF	Polychlorinated fluorenes		

Although the isotope patterns differ depending on the number of chlorine atoms, an unambigious identification of specific congeners might become difficult due to co-elution of various contaminants from the capillary column. Hence, a highly sophisticated clean-up in combination with high resolution GC-MS is required to minimize the possibilty of false positive results due to the similar structure of many of these environmental pollutants.

5 Summary

Within the last two decades detection limits in residue analysis were lowered by several orders of magnitude. Consequently, compounds which have contaminated the environment for years can nowadays be identified and determined in biological samples and foodstuffs at low ppt- and even ppq-levels.

Use of capillary columns of different polarity generally provides an unambigious separation of highly toxic compounds from less toxic ones. This is of great importance because a meaningful risk assessment can only be based on a congener-specific analysis of contaminants in food and other biological samples.

However, high sensitivity attainable with modern instruments and similar physico-chemical properties of various compound classes demand not only a discriminating sample clean-up and a comprehensive quality assurance program but also a high degree of critical faculty from the analyst with respect to the results.

Acknowledgement: The author wishes to thank Dr. J. J. RYAN, Health and Welfare Canada, for critically reading the manuscript and for fruitful discussions.

References

1. STALLING, D. L.,TINDLE, R. C., JOHNSON, D. L. (1972) J. Assoc. Off. Anal. Chem. **55**: 32–38

2. MURPHY, P. G. (1972) J. Assoc. Off. Anal. Chem. **55**: 1360

3. SMREK, A. L., NEEDHAM, L. L. (1982) Bull. Environm. Contam. Toxicol. **28**: 718–722

4. BECKER, G. (1982) Organohalogen-, Organophosphor- und Triazin-Verbindungen, DFG-Methodensammlung, Methode S 8

5. Multiple Residue Methods for Organophosphorus Pesticides (1984), Carbon Column Clean-up Method. AOAC-Method 29.054

6. SMITH, L. M. (1981) Anal. Chem. **53**: 2152-2154

7. SMITH, L. M., STALLING, D. L., JOHNSON, J. L. (1984) Anal. Chem. **56**: 1830–1842

8. BECK, H., ECKART, K., KELLERT, M., MATHAR, W., RÜHL, Ch.-S., WITTKOWSKI, R. (1987) Chemosphere **16**: 1977–1982

9. FÜRST, P., KRÜGER, C., MEEMKEN, H.-A., GROEBEL, W. (1989) Chemosphere **18**: 439–444

10. STANLEY, J. S., BOGGESS, K. E., ONSTOT, J., SACK, T. M., REMMERS, J. C., BREEN, J., KUTZ, F. W., CARRA, J., ROBINSON, P., MACK, G. A. (1986) Chemosphere **15**: 1605

11. KANNAN, N., TANABE, S., WAKIMOTO, T., TATSUKAWA, R. (1987) Chemosphere **16**: 1631–1634

12. KRÜGER, C. (1988) PhD Thesis, University of Münster, West Germany

13. HUPFELD, M., FÜRST, P., GROEBEL, W. (1988) GC-MS-Bestimmung von Benzo(a)pyren in geräucherten Fleischerzeugnissen nach Reinigung an Carbopack C, Regionaltagung der Fachgruppe "Lebensmittelchemie und gerichtliche Chemie" in der GDCh, Bonn, West Germany

14. WHO Consultation on Quality Control Studies on Levels of PCB, PCDD and PCDF in Human Milk (1987) WHO Regional Office for Europe, Copenhagen (IPC/CEH 541/E)

15. LINDSTRÖM, G. (1988) PhD Thesis, University of Umea, Sweden

16. CLEMENT, R. E., BOBBIE, B., TAGUCHI, V. (1986) Chemosphere **15**: 1147–1156

17. SCHUBERT, R. (1985) GIT Fachz. Lab. **29**: 1175–1177

18. LORENZ, H., NEUMEIER, G. (1983) Polychlorierte Biphenyle. Gemeinsamer Bericht des Bundesgesundheitsamtes und des Umweltbundesamtes, MMV Medizin Verlag, München

19. HUTZINGER, O., SAFE, S., ZITKO, V. (1974) "The Chemistry of PCBs", CRC Press, Cleveland, Ohio

20. BUSER, H. R., BOSSHARDT, H. P., RAPPE, C. (1978) Chemosphere **7**: 109–119

21. RAPPE, C., MARKLUND, S., BERGQVIST, P.-A., HANSSON, M. (1982) Chemica Scripta **20**: 56–61

22. SAFE, S., SAFE, L., MULLIN, M. D. (1987) in "Environmental Toxin Series 1: Polychlorinated Biphenyls (PCBs): Mammalian and Environmental Toxicology" (Safe, S., Hutzinger, O., Eds.), Springer-Verlag, Berlin–Heidelberg

23. SAFE, S. (1984) CRC Crit. Rev. Toxicol. **13**: 319

24. PARKINSON, A., SAFE, S. (1987) in "Environmental Toxin Series 1: Polychlorinated Biphenyls (PCBs): Mammalian and Environmental Toxicology" (Safe, S., Hutzinger, O., Eds.), Springer-Verlag, Berlin–Heidelberg

25. YOSHIMURA, H., YOSHIHARA, S., OZAWA, N., MIKI, M. (1979) Ann. N. Y. Acad. Sci. **320**: 179

26. POLAND, A., KNUTSON, J. C. (1982) Ann. Rev. Pharmacol. Toxicol. **22**: 517

27. Verordnung über Höchstmengen an Schadstoffen in Lebensmitteln vom 23. 3. 1988 (BGBl I S. 422)

28. BALLSCHMITER, K., ZELL. M. (1980) Fresenius Z. Anal. Chem. **302**: 20

29. MES, J. (1988) Personal communication

30. SAFE, S., SAFE, L., MULLIN, M. (1985) J. Agric. Food. Chem. **33**: 24

31. SCHULTE, E., MALISCH, R. (1984) Fresenius Z. Anal. Chem. **319**: 54–59

32. MES, J., TURTON, D., DAVIES, D., SUN, W. F., LAU, P. J., WEBER, D. (1987) Intern. J. Environ. Anal. Chem. **28**: 197–205

33. FÜRST, P. (1987) Unpublished data

34. SCHULTE, E., MALISCH, R. (1983) Fresenius Z. Anal. Chem. **314**: 545–551

35. ZOLLER, W., SCHÄFER, W., CLASS, T., BALLSCHMITER, K. (1985) Fresenius Z. Anal. Chem. **321**: 247

36. MULLIN, M. D., POCHINI, C. M., MC CRINDLE, S., ROMKES, M., SAFE, S., SAFE, L. (1984) Environ. Sci. Technol. **18**: 468

37. FÜRST, P., KRÜGER, C., MEEMKEN, H.-A., GROEBEL, W. (1987) J. Chromatogr. **405**: 311–317

38. FÜRST, P., KRÜGER, C., MEEMKEN, H.-A., GROEBEL, W. (1987) Z. Lebensm. Unters. Forsch. **185**: 394–397

39. RÖNNEFAHRT, B. (1987) Dtsch. Lebensm. Rundsch. **83**: 214–218

40. FÜRST, P. (1988) Spectra **11 (2)**: 26–29 ·

41. THOMA, H., RIST, S., HAUSCHULZ, G., HUTZINGER, O. (1986) Chemosphere **15**: 649

42. DE KOK, J. J., DE KOK, A., BRINKMAN, U. A. Th. (1979) J. Chromatogr. **171**: 269

43. LIEPINS, R., PEARCE, E. M. (1976) Environ. Health Perspect. **17**: 55

44. FRIES, G. F. (1985) Critical Reviews in Toxicology **16**: 105–156

45. FEHRINGER, N. V. (1975) J. Assoc. Off. Anal. Chem. **58**: 978

46. FEHRINGER, N. V. (1975) J. Assoc. Off. Anal. Chem. **58**: 1206

47. EYSTER, J. T., HUMPHREY, H. E. B., KIMBROUGH, R. (1983) Arch. Environ. Health **38**: 47

48. SUNDSTRÖM, G., HUTZINGER, O., SAFE, S. (1976) Chemosphere **5**: 11

49. AUST, S. D., MILLIS, C. D., HOLCOMB, L. (1987) Arch. Toxicol. **60**: 229–237

50. ROBERTSON, L. W., SAFE, S. H., PARKINSON, A., PELLIZZARI, E., POCHINI, C., MULLIN, M. D. (1984) J. Agric. Food Chem. **32**: 1107–1111

51. RAPPE, C. (1984) Environ. Sci. Technol. **18**: 78 A

52. SCHOLZ, B., ENGLER, M. (1987) Chemosphere **16**: 1829

53. HAGENMAIER, H., BRUNNER, H. (1987) Chemosphere **16**: 1759

54. BUSER, R., BOSSHARDT, H. P. (1976) J. Assoc. Off. Anal. Chem. **59**: 562

55. BUSER, R., BOSSHARDT, H. P. (1978) Mitt. Geb. Lebensm. Hyg. **69**: 191

56. SWANSON, S. E., RAPPE, C., KRINGSTAD, K. P., MALMSTROM, J. (1989) Abstract SE 01, Dioxin 87, Las Vegas

57. KUEHL, D. W., BUTTERWORTH, B. C., DE VITA, W. M., SAUER, C. P. (1987) Biomed. Environ. Mass. Spectr. **14**: 443

58. BECK, H., ECKART, K., MATHAR, W., WITTKOWSKI, R. (1988) Chemosphere **17**: 51

59. MARKLUND, S., RAPPE, C., TYSKLIND, M., EGEBAECK, K.-E. (1987) Chemosphere **16**: 29

60. BALLSCHMITER, K., BUCHERT, H., NIEMCZYK, R., KNUDER, A., SWEREV, M. (1986) Chemosphere **15**: 901

61. RYAN, J. J., LIZOTTE, R., LAU, B. P.-Y. (1985) Chemosphere **14**: 697

62. BAUGHMAN, R., MESELSON, M. (1973) Environ. Health Perspect. **5**: 27–35

63. GROSS, M., LAY, J. O., LYON, P. A., LIPPSTREU, D., KANGAS, N., HARLESS, R. L., TAYLOR, S. E., DUPUY, A. E. (1984) Environ. Res. **33**: 261

64. NYGREN, M., RAPPE, C., LINDSTRÖM, G., HANSSON, M., BERGQVIST, P.-A., MARKLUND, S., DOMEL-LOF, L., HARDELL, L., OLSEN, M. (1986) in "Chlorinated Dibenzodioxins and Dibenzofurans in the Total Environment" (Rappe, C., Choudhary, G., Keith, L., Eds), Lewis Publishers, Vol. III

65. PATTERSON, D. G., HAMPTON, L., LAPEZA, C. R., BELSER, W. T., GREEN, V., ALEXANDER, L., NEED-HAM, L. L. (1987) Anal. Chem. **59**: 2000

66. RAPPE, C., BUSER, R., STALLING, D. L., SMITH, L. M., DOUGHERTY, R. C. (1981) Nature **292**: 524

67. MITCHUM, R. K., MOLER, G. F., KORFMACHER, W. A. (1980) Anal. Chem. **52**: 2278

68. RAPPE, C. (1985) WHO-Consultation on Organohalogen Compounds in Human Milk and Related Compounds, Bilthoven

69. FÜRST, P., MEEMKEN, H.-A., GROEBEL, W. (1986) Chemosphere **15**: 1977

70. VAN DEN BERG, M. VAN DER WIELEN, F. M. W., OLIE, K. (1986) Chemosphere **15**: 683

71. ONO, M., KASHIMA, Y., WAKIMOTO, T., TATSUKAWA, R. (1987) Chemosphere **16**: 1823

72. RAPPE, C., ANDERSON, R., BERGQVIST, P.-A., BROHEDE, C., HANSSON, M., KJELLER, L.-O., LINDSTRÖM, G., MARKLUND, S., NYGREN, M., SWANSON, S. E., TYSKLIND, M., WIBERG, K. (1987) Chemosphere **16**: 1603

73. RYAN, J. J. (1988) Personal communication

74. SWEREV, M., BALLSCHMITER, K. (1987) J. High Res. Chromatogr. **10**: 544

75. STALLING, D. L., SMITH, L. M., PETTY, J. D., HOGAN, J. W., JOHNSON, J. L., RAPPE, C., BUSER, H.-R. (1983) in: "Human and Environmental Risks of Chlorinated Dioxins and Related Compounds" (Tucker, R. E., Young, A. L., Gray, A. P., Eds.), Plenum Publishing Corporation, p. 221

76. RAPPE, C., NYGREN, M., LINDSTRÖM, G., BUSER, H. R., BLASER, O., WUTHRICH, C. (1987) Environ. Sci. Technol. **21**: 964

77. BECK, H., ECKART, K., MATHAR, W., WITTKOWSKI, R. (1989) Chemosphere **18**: 417–424

78. FÜRST, P. (1987) Unpublished data

79. KIMBROUGH, R. D. (Ed.) (1980) "Halogenated Biphenyls, Terphenyls, Naphthalenes, Dibenzodioxins and Related Products", Elsevier, Amsterdam

80. HASS, J. R., MC CONNELL, E. E., HARVAN, D. J. (1978) J. Agric. Food Chem. **26**: 94

81. WILLIAMS, C. H., PRESCOTT, C. L., STEWART, P. B., CHOUDHARY, G. (1985) in "Chlorinated Dioxins and Dibenzofurans in the Total Environment II" (Keith, L. H., Rappe, C., Choudhary, C., Eds.), Butterworth Publishers, Stoneham

82. RORDORF, B. F., FREEMAN, R. A., SCHROY, J. M., GLASGOW, D. G. (1986) Chemosphere **15**: 2069

83. PAASIVIRTA, J., TARHANEN, J., JUVONEN, B., VUORINEN, P. (1987) Chemosphere **16**: 1787

III.3 Headspace Gas Chromatography of Highly Volatile Compounds

S. Vieths

1 Introduction

A combination of methods and technological procedures used to obtain information about the composition, nature or state of liquid and solid bodies by analysis of their surrounding gas phase is called headspace analysis [1].

Headspace gas chromatography (HSGC) is one of the most elegant methods of gas chromatographic analysis. Sample clean-up or preparation, normally necessary in food analysis, can either be totally eliminated or at least can be reduced in many cases.

In the case of a solid or liquid sample in thermodynamic equilibrium with the surrounding gas phase, we are talking about *equilibrium or static HSGC*. If the sample is extracted by a continuous gas flow, it is termed a *dynamic headspace (HS)-method.*

In static HS-processes inert gas is injected into the GC instead of a solvent. This results in the fact that no disturbing solvent peak occurs in the chromatogram, provided that no volatile solvent has been added to the sample itself. In addition, no contamination of the analytical column by non-volatile components from the sample matrix occurs, which otherwise quite often cannot be avoided if liquid extracts of food samples are injected. The same is also true for dynamic HS methods.

Since static HS techniques are easily automated and the results are well reproducible, they are more and more used as rapid methods as for example in quality control [2].

This paper describes possible applications of HSGC-methods in the field of residue analysis in foods and packaging materials, and clarifies this by means of selected examples. The theoretical basis is briefly discussed as well. Particular importance is attached to the description of technical possibilities and practical procedures. The selected examples of application described in Chapter III.3.–5 are to clarify the possible applications of HSGC and simplify the familiarization with the special literature, however, they do not represent a complete evaluation of the literature. In conclusion, the different techniques are compared, and advantages and disadvantages are discussed.

2 Equilibrium Headspace Gas Chromatography

2.1 Theory

Static HSGC makes use of the thermodynamic equilibrium between the volatile components of a liquid or solid sample and the surrounding gas phase in a closed system.

The partial vapor pressure pi of a component i above a liquid sample is expressed as

$$p_i = p^o_i X_i A_i \text{ (real system)} \tag{1}$$

where p^o_i is the vapor pressure of the pure component i at a defined temperature. X_i is the mole fraction of component i in the liquid and A_i is the activity coefficient. In the ideal system the activity coefficient is 1 and the equation reduces to Raoults law:

$$p_i = p^o_i X_i \tag{2}$$

The partial pressure p_i is proportional to the measured peak area F_i:

$$F_i = C_i p_i \tag{3}$$

Among other parameters the component specific constant C_i depends on the detector response.

In the quantitative equilibrium HSGC the mole fraction of the component i in the liquid phase is to be determined. The mole fraction X_i is obtained by combining the equations (1) and (3):

$$X_i = \frac{F_i}{C_i p^o_i A_i} \tag{4}$$

The relationship between the peak area and the mole fraction has to be determined experimentally by calibration. Calibrations, however, especially if they cover very different concentration ranges do not always show linear relationships because the activity coefficient can be a function of the mole fraction. Nevertheless, if HSGC is applied in trace analysis this effect usually is insignificant and ordinarily linear calibration graphs are obtained.

Another important prerequisite in HSGC is constant temperature, as the vapor pressure changes with the temperature in accordance with the Clausius-Clapeyron equation:

$$\frac{d(\ln p)}{dT} = \frac{H_v}{RT^2} \tag{5}$$

where: H_v = vaporization enthalphy, R = gas constant, T = temperature

In summary, it can be stated that in a closed system there is a quantitative relationship between the concentration of a compound in a liquid or solid phase and its partial pressure in the surrounding gas volume. In quantitative HSGC this relationship is determined experimentally be means of a calibration. For this, it is necessary to precisely control the constancy of temperature. Constant thermostatting times or thermostatting until equilibrium is reached are therefore required [3–4].

2.2 Increasing Analytical Sensitivity

Apart from instrumental parameters and the compound-specific value of the detector response the analytical sensitivity of static HSGC methods depends on the real partition coefficient between liquid phase and the surrounding gas volume. Therefore, it is a function of the sample matrix, resulting in different detection limits if a specific compound is measured in different matrices. For example, non-polar compounds can be determined in polar matrices with a lower detection limit than in non-polar matrices. An example for that would be the determination of volatile halocarbons (VHCs) from aqueous matrix compared to the determination of these compounds in fat. Naturally there is an opposite situation in the case of polar analytes.

Analytical sensitivity is increased if a greater amount of the analyte is transferred to the column within one GC analysis. This can be accomplished by the methods described below:

(a) Increasing the vapor pressure by raising the temperature. This increases the share of volatile compounds in the gas phase. The increase of sensitivity by this method, however, is not drastic and is limited because with high temperatures chemical interactions as well as too high pressures in the HS-vessel can occur.

(b) By addition of an electrolyte to aqueous solutions, or water to organic solutions the activity coefficients of compounds in the liquid phase can be increased, leading to higher vapor phase concentrations. Figure III.3.–1 illustrates this effect: The addition of water to a solution of styrene, butyl acrylate, acrylonitrile, and n-butanol in DMF increases the peak area obtained in HS analysis considerably. The greatest effect is obtained in the case of non-polar styrene while polar n-butanol shows the smallest change [3].

(c) Injecting large gas volumes also increases the analytical sensitivity. If capillary columns are used this often leads to the problem that the injection of large gas volumes requires long sampling times, resulting in band broadening which in turn can cause peak distortion and poor resolution, especially in the case of highly volatile compounds [4].

Figure III.3.–2 illustrates this effect: A cheese sample was analyzed for its flavor compounds. Chromatogram **A** was obtained using conventional HS-sampling with splitless injection, resulting in sharp peaks. Sampling time was 4.8 seconds. However, when sampling time was increased to 24 seconds a significant deterioration of the early peaks occurred (chromatogram **B**). The latter peaks corresponding to the fatty acids were not affected by the long sampling time, because their capacity factor was sufficiently high. Their peak heights therefore increased proportionally to sampling time.

By means of cryogenic focusing the observed effect can be avoided. For this the first part of the column is chilled during the injection, for example by inserting the first loop of the column into a Dewar vessel containing liquid nitrogen. This reduces the migration rate

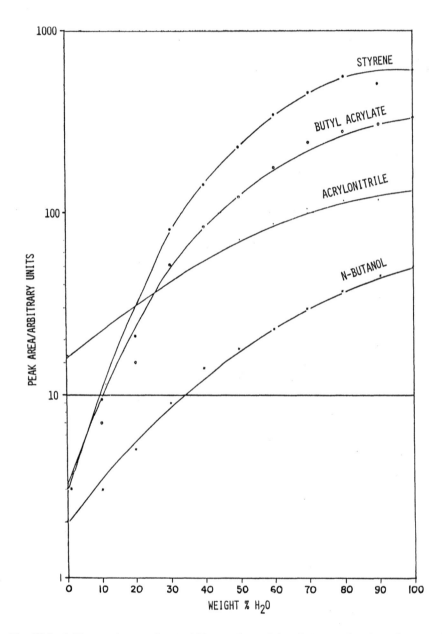

Fig. III.3.–1 The peak area due to 120 ppm by weight of styrene, butyl acrylate, acrylonitrile and n-butanol in DMF as a function of the the water content of the mixture [3]

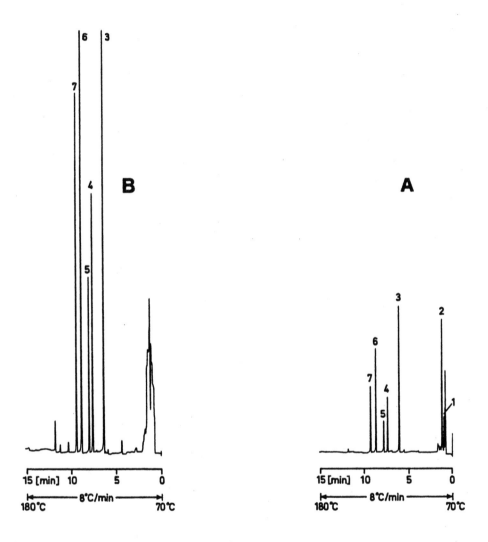

Fig. III.3.–2 Analysis of flavor compounds present in cheese, utilizing headsace sampling,
Column: 25 m × 0,32 mm i. d. capillary column coated with free fatty acid
phase; film thickness 1 μm. *Column temperature:* programmed as indicated.
Sample: 2 g ground cheese. *Headspace equilibration temperature and time:*
90 °C, 60 min. *Sampling time:* **A** 4,8 s, **B** 24 s. *Identified peaks:* **1** acetaldehyde, **2** ethanol, **3** acetic acid, **4** propionic acid (130 ppm), **5** isobutyric
acid, **6** butyric acid, **7** isovaleric acid (85 ppm) [5]

of the compounds and consequently their peak widths. An arrangement for manual injection of great HS-volumes is depicted in Figure II.3.–15 [6].

Some commercially available injection systems allow cryofocusing. Usually, they are designed to inject volatile compounds thermally desorbed from Tenax or XAD (see below). Slightly modified they may also be used for manual injections. Figure III.3.–3 shows a modified thermal desorption cold trap injector. A short piece of fused silica capillary column which can be cooled with liquid nitrogen is installed as the cold trap. To transfer the focused compounds to the GC, the trap is heated up rapidly [7]. Using cryogenic focusing, manual injection of several ml HS-gas onto capillary columns without affecting the shape of the peaks formed can be performed. In the case of purge and trap methods (discussed below) such cold traps are used to focus volatile compounds from gas volumes of more than 100 ml.

2.3 Performing Equilibrium HSGC

Equilibrium HSGC requires gas tight sample vessels, the capability to maintain a constant temperature, and to withdraw reproducible samples from the vapor phase. This can be accomplished either manually with gas tight syringes, or by use of an automatic dosing system. Since it is common that in residue analysis a large number of samples have to be examined for definite contaminants, HS-autosamplers are used very frequently. Therefore, this chapter mainly describes the use of the automated HSGC.

The use of gas tight syringes causes several problems: The temperature of the syringe has to be above the sample temperature otherwise condensation on the cylinder or the piston can occur. Therefore the syringe has to be thermally controlled which might cause handling problems. The syringe also has to be thoroughly cleaned at reduced pressure and increased temperature. Furthermore, the column head pressure has to be as low as possible to allow a fast injection and avoid a possible snapping of the piston [3]. If capillary columns are used, this is less problematic because of the lower initial column pressure. Manual quantitative analysis has to be performed utilizing an internal standard to compensate for the relatively poor reproducibility of the injection.

In HS autosamplers samples are placed into septum-sealed glass vessels. Before analysis, both, samples and standards are thermostatted either for a constant time or until thermodynamic equilibrium is reached. Figure III.3.–4 shows the principle of time- and pressure-controlled sampling, as applied in an automated headspace device. After the needle has pierced through the septum, carrier gas flows into the vial until the pressure equals that at the head of the column. The magnetic valve V_1 then interrupts the carrier gas flow for a few seconds. This results in the gas expanding through the needle and thus the volatiles are transferred onto the column. After the injection the valve opens the carrier gas flow stopping the injection by repressurizing the vessel [4]. The injected HS volume can be regulated with the sampling time.

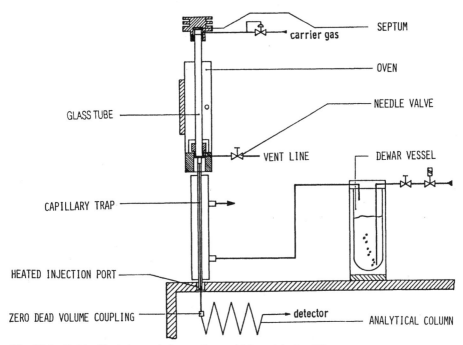

Fig. III.3.–3 Modified thermal desorption cold trap injector [7]

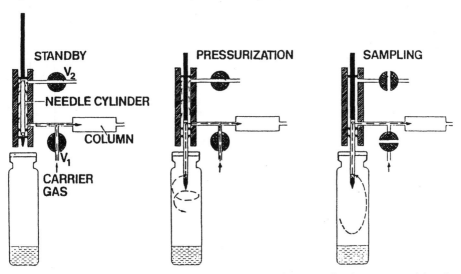

Fig. III.3.–4 Principle of the time and pressure-controlled sampling in automated head-space analysis [4]

Several aspects of using such pneumatic HS autosamplers in combination with capillary columns are discussed by ETTRE et al. [8]. In principle the split as well as the splitless injection can be carried out. The problem, however, with this kind of pneumatic sampling is that double sampling might occur if the pressure in the vessel after the thermostatting period is higher than the initial pressure at the column head. This may occur if for example a volatile solvent was added to the sample and a high temperature setting is used. By setting a small split ratio double sampling can be avoided, because the split method generally requires a higher column head pressure to maintain the optimal carrier gas flow. The use of capillary columns with an internal diameter of 0.53 mm or more also requires a higher initial pressure and therefore can be useful in specific applications.

Another automated system is based on a different principle, the so-called valve and loop-technique in which a changeable sample loop is used for dosing the HS gas [9−10]. Figure III.3.−5 shows a schematic diagram of such a device.

The sampling procedure also begins with a pressurization step following a thermostatting period. Then valve V_1 is opened and the HS gas fills up the metering loop. By operating the valve V_1 once more, the carrier gas flow is switched onto the loop and the sample is transferred to the GC-column. If a capillary column is used, it is recommended to work with a small split ratio or to use a cryofocusing step. However, direct splitless injection,

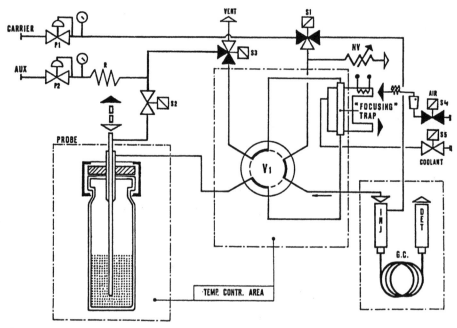

Fig. III.3.−5 Pneumatic circuit of a valve and loop headspace autosampler [9]

as it is used in time and pressure-controlled sampling cannot be applied in this system [9–10]. The split method generally does not cause a lower analytical sensitivity because the limiting factor here is band broadening caused by long injection times. With the split-less method a shorter sampling time has to be chosen or smaller gas volume has to be injected. Therefore, the maximum gas volume which can be dosed onto the GC-column without influencing the shape of the peaks is similar in both methods.

HS autosamplers, which work by the valve and loop-mechanism can be easily combined with any type of gas chromatograph, however, this is not true with time and pressure controlled sampling devices.

Like in manual injection, other automated systems make use of gas tight syringes.

The instrumental precision of the pneumatic HS injection is very high. In general, variation coefficients are below 1 % when analyzing aqueous solutions. As an example for the good reproducibility of the automated HS injection, Figure III.3.–6 shows the gas chroma-tograms obtained from repeated HS analyses of a 0.01 % solution of acetone in water [11]. However, in food analysis samples of very complex compositions have to be analy-zed which often leads to significantly higher variation coefficients.

2.4 Quantitation

Before discussing the practical aspects of quantitative determinations with the equilibrium HSGC, attention has to be focused to HSGC applications without need for exact quanti-tation, for example the control of production or fermentation processes or if end products have to be screened for residues of technological auxiliary substances in a short time. Aroma analysis (quality control) is another example that can be cited here [12]. Very good reproducibility of a method allows us to evaluate measurements with experience to inter-pret the results. However, this article will not discuss such applications but will explore the quantitative aspects of HSGC.

2.4.1 External Calibration

In order to obtain the concentration of a component i in a food sample from the peak area of an equilibrium HS gas chromatogram both, the instrumental parameters and the com-pound specific detector response as well as the real partition coefficient have to be con-sidered in the calibration. If the partition coefficient is known or a complete transition of the analyte from the liquid or solid sample into the HS gas is assumed, the calibration can be accomplished by analyzing test gases of a known amount. This method was mainly used in the beginning of the utilization of HSGC in food analysis in combination with manual HS sampling [15]. But in general, matrix effects have to be considered in the cali-bration. Therefore, the composition of the sample under investigation and the calibration standards must be comparable. A prerequisite for this is that a non-contaminated matrix is available. In contrast to manual injections the use of an autosampler does not necessarily require internal standards because of its high precision. Nevertheless, it can be helpful in finding leaks in HS-vessels or facilitate the discovery of sources of trouble during sample preparation (e. g. extraction or derivatisation).

Examples for this important calibration method are mentioned at the end of this Chapter (III. 3.5.1.1, III.3.5.3, III.3.5.4, III.3.5.8, III.3.5.9).

2.4.2 Standard Addition Method

If no matrix without analyte residues for producing a calibration series is available and it is not possible to simulate the matrix effect by mixing the main constituents of the sample, calibration can be achieved with the sample itself. For this, as in external calibration, different amounts of the analyte are added and then the samples are analyzed with and without standard addition. The concentration in the original sample is determined on the basis of the proportionality of the increase in peak area. For evaluation the peak area or peak area ratio is plotted versus the added amount. The concentration of the analyte in the sample is determined from the intersection of the linear calibration graph and the x-axis. In principle all determinations mentioned above can be achieved using the standard addition method. If this procedure is used, every single sample requires its own calibration causing drastic increases in the number of measurements. However, the advantage of this method is that more accurate results are obtained because all systematic mistakes, including matrix effects, are compensated.

The amount of work can be reduced, if only one sample as well as one sample containing one standard concentration are analyzed, however, this results in a loss of accuracy. The standard addition method is often used to confirm results that have been obtained by external calibration.

If a great number of determinations of an analyte from the same type of sample has to be performed the method of choice is external calibration. Only occasionally or in special cases the results are confirmed by standard addition (see also III.3.5.1.2).

Finally, some general remarks about quantitative determination of residues and contaminants by equilibrium-HSGC: the precision of the results highly depends on the homogeneity of the samples. Homogenisation of food, if examined for volatile compounds, must be carried out carefully to avoid losses of the analyte. Since in automated HSGC generally only small sample weights are analyzed, the homogeneity of the sample itself or the reproducibility of the sample preparation often limits the precision of the results.

The results of a study to elaborate a standard method to determine volatile halocarbons in different foodstuffs illustrate this particular matrix problem. During the analysis of vegetable oil with the standard addition method variation coefficients between 5 % and 7 %

were obtained. The same method showed variation coefficients up to 13 % in the analysis of bacon [17]. In spite of homogenisation there were different amounts of fat and fibre in these samples. The effect described is explainable and can hardly be avoided because volatile halocarbons are mainly found in the fat compartment of food. Similar results have been reported from another study: in the analysis of food composites for volatile halocarbons, good quantitative results were obtained analyzing oil/fat and also diaries, while in the analysis of meat and beverages only qualitative evaluations were possible [18].

In addition to the quantitation methods discussed here the Multiple Headspace Extraction (MHE) which under certain circumstances allows a matrix-independend quantitation, is introduced and discussed in Chapter III.3.4.

3 Dynamic Headspace Gas Chromatography

Dynamic HS methods are those techniques in which a continuous gas flow is used to carry out the extraction of the sample. Some authors refer to "dynamic HS" when the gas flow is led over the sample and of "strip and trap" or "purge and trap", when the extraction gas is passed through a liquid sample or through a suspension of a solid sample. In quantitative food analysis purge and trap methods are most important. This method was originally developed to determine volatile halocarbons in water [19]. Another dynamic HS method is the so-called closed loop strip/trap-technique in which volatile compounds from a condensed phase are trapped on a sorbent by pumping the headspace gas in a closed circuit via the trap and the condensed phase. This method originally was developed by GROB and GROB [20−21]. All dynamic HS techniques require an enrichment step (trap) prior to GC analysis.

This Chapter describes methods of dynamic HS analysis. Selected examples from the field of quantitative analysis of foods for residues and contaminants are to be found in Chapter III.3.5. The description is confined to those devices in which the extraction gas can be led through the liquid samples.

3.1 Theory

In the case of a liquid sample the concentration C_G of a volatile analyte in the gas phase can be calculated according to equation **(6)** [1]:

$$C_G = \frac{C^o_L}{K} e^{-\frac{V_G}{KV_L}} \tag{6}$$

where: C^o_L = initial concentration of the analyte in the liquid phase
        K = partition coefficient
        V_G = purge gas volume
        V_L = volume of the liquid sample

A similar relationship has also been described by DROZD and NOVAK [22]. However, these expressions only hold in the case of thermodynamic equilibrium, homogeneity of the sample and the purge gas, and constant sample temperature and purge flow rate. A slightly modified equation which considers the gas volume above the liquid sample, was introduced by NOIJ et al. [23].

$$R = 1 - e^{\left(\dfrac{-F(t - t_d)}{K V_l + V_g} \right)} \tag{7}$$

where: R = recovery, F = purge gas flow rate, t = purge time, t_d = gas hold-up time in the glass tube, K = mass based gas-liquid distribution constant, V_l = sample volume, V_g = gas volume over the liquid sample in the sample flask.

The theoretical basis for the closed loop strip/trap-method is described by NOVAK et al. [24].

The large number of dynamic HS methods developed results in many small differences in theoretical considerations, which will not be described here.

3.2   Performing Dynamic HSGC

In all dynamic HS methods the sample is extracted by a continuous stream of gas. Sometimes more than 1000 ml of purge gas are needed for the extraction of one sample [25] and therefore, the concentration of the analyte has to be increased prior to the GC analysis. This can be accomplished by solvent absorption, adsorption on solid sorbents such as activated charcoal or macroporous polymers such as Tenax- or XAD-resins, or by cryofocusing in cold traps. Since the solid sorbents have different sampling characteristics and the desorption rate (thermodesorption or solvent elution) depends on the polarity and the boiling point of the analyte, both have to be optimized for every specific analytical problem.

Solid sorbents behave like chromatographic columns. After a specific retention the enriched compounds are emitted from the sorbent. In practice, the adsorption of polar volatiles on organic resins causes problems. In principle, this is also true in the case of cold traps which use capillary columns as sorbents.

If activated charcoal is used, elution with a solvent is often necessary to desorb the trapped compounds. For example, non-polar compounds are not easily desorbed from activated charcoal if thermal desorption is applied. In this case very long desorption times at high temperatures are necessary which can cause chemical interactions resulting in losses of analyte or formation of artifacts. Carbon disulphide is known to be a good eluant. Thermal desorption with activated charcoal as a sorbent is possible, if it is employed in combination with organic resins. For example, if the trap consists of a combination of XAD or Tenax and activated charcoal and if the stripping gas passes through the organic resin first, the non-polar compounds and some of the polar volatiles are collected in this portion. On the coal only the unadsorbed fraction of the polar volatile compounds is enriched. These particular compounds are thermally desorbable without excessive desorption times.

In many cases the use of organic polymers is also combined with the elution method. This leads to a loss of analytical sensitivity because only an aliquot of the trapped compounds is transferred onto the GC column. The use of thermal desorption avoids this because the entire amount of the trapped compounds can be analyzed in one GC run. If desorption is accomplished from Tenax or XAD-tubes onto capillary columns, a special instrumental problem occurs: desorption requires several minutes and carrier gas flow is 5–10 orders of magnitude higher than the normal carrier gas flow through capillary columns. Therefore, before the volatiles are transferred to the GC, an additional sorption/desorption step has to be included. Here mostly the cryofocusing/thermal desorption technique is employed. Thermal desorption cold trap injectors which have been designed especially for this use, are commercially available [7]. The use of capillary columns with a particularly large internal diameter (0.53 or 0.75 mm) makes the second sorption/desorption step in some cases avoidable [26].

In general, thermal desorption steps involve the risk of decomposition or artifact formation.

Purge and trap equipment has been developed for off-line as well as for on-line use. The off-line use can be combined with a solvent as the trapping material as well as with adsorption of the volatiles on activated charcoal or organic resins. For thermal desorption of off-line loaded adsorption tubes autosamplers are available [4]. In the on-line use of purge and trap-devices organic sorbents as well as cryofocusing can be applied.

Dynamic HS techniques, especially those in residue analysis particularly often used purge and trap methods, have been carried out with a number of self-constructed equipment. Today, different automated devices are commercially available, two of which shall be described here. Figure III.3.–7 shows a schematic diagram of an on line purge and cold trap-injector (PTI) in which cryofocusing and thermal desorption are applied.

A flow of purge gas (helium, also carrier gas) is passed through the sample by means of a glass frit 2, which ensures a highly dispersed purge flow. The sample flask 1 can be thermostatted using an oven which is not shown in the illustration. The flow, containing both volatiles and water vapor, passes condenser 3 which is kept at – 15 °C. In this trap most of the water vapor is frozen and eliminated in order to avoid blockage of the second cold trap. After passing a glass tube in a heated compartment 4, the volatiles are trapped in a small piece of fused silica tubing 5, coated with SP SIL 5CB and held at a minimum of – 120 °C. This low temperature is maintained by an air stream cooled with liquid nitrogen from a Dewar vessel 6. The capillary trap is connected to the analytical GC capillary column 7. For thermal desorption, the cold trap can then be heated up to e. g. 200 °C

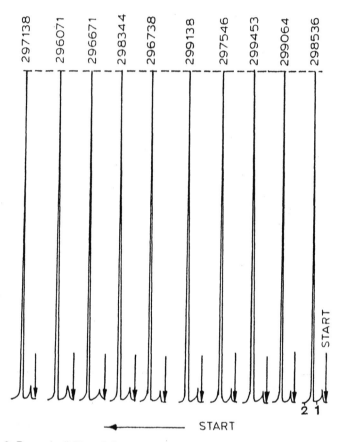

Fig. III.3.–6 Reproducibility of the pneumatic injection into the gas chromatograph of acetone vapor above its water solution [1]

within seconds. Purge, cryofocusing, thermal desorption and GC analysis are programmable, however, with this particular system the samples have to be changed manually [7, 27]. Condenser 3, fixed before the cold trap effectively removes the water from the purge gas, but also removes volatile compounds with higher boiling points, for example hydrocarbons with more than 11 C-atoms are held back and therefore cannot be analyzed. In contrast, other equipment in which organic polymers are used as sorbents allows to determine C-14 and C-15 hydrocarbons from aqueous matrices [23].

The pneumatic circuit of a sequential purge and trap autosampler which is commercially available is shown schematically in Figure III.3.–8.

In this system, up to 24 septum-sealed vessels, containing the samples, can be thermally controlled in a carousel heated by a silicon oil bath. After a preset heating time, the septum is pierced by a twin-needle, through which the purge gas, is led in and out. The gas which contains extracted volatiles passes a small amount of a sorbent. The trapped volatiles are then transferred to the GC column by thermal desorption. If capillary columns are used, a small splitting (1:10) must be provided or a cryofocusing step performed [9, 28]. In the standard equipment of this instrument the latter is not included.

The great advantage of this device is that up to 24 samples can be sequentially examined, so that it can be used in the same way as an equilibrium HSGC-autosampler. On the other hand, the contact between the purge gas and the liquid sample is considerably more intense with the glass frit in the PTI system. Gas supplied through a needle represents a disadvantage because purge and trap-methods in residue analysis are carried out

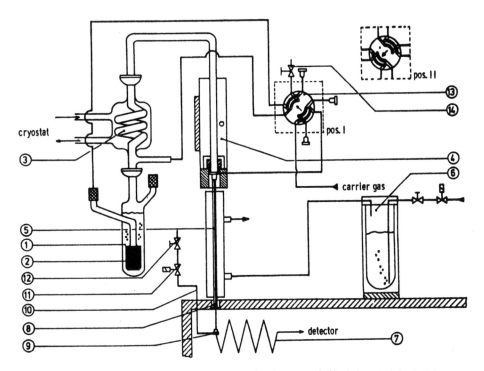

Fig. III.3.–7 Schematic diagram of an automated purge and cold trap injector [7]
1 sample flask, **2** glass frit, **3** condenser, **4** heated compartment, **5** cold trap, **6** Dewar vessel, **7** analytical capillary column, **8** rod, **9** hexagonal nut, **10** vent line, **11** solenoid valve, **12** needle valve, **13** switching valve, **14** constriction

until total extraction if possible. Therefore, in many self-designed purge and trap devices attempts have been made to ensure intimate contact between the purge gas and the sample by either vigorous stirring or use of large, small-porous frits [25, 30–31]. In the case of incomplete extraction, the calibration matrix effect has to be taken into account (see also III.3.2.4.1). On the other hand, an advantage of the described sequential purge and trap analyzer is that the Multiple Headspace Extraction can be carried out according to a dynamic HS technique (see below) [29]. Compared to the PTI system whose possibilities of application have already been extensively examined [7, 23, 27, 32], the sequential analyzer seems to have had little use up to now. One example of its use, the determination of VC-monomer in PVC, is described in Chapter III.3.4 [29].

Finally, another interesting variant of a dynamic HS technique is presented – the closed loop strip/trap-method, in which volatile compounds from a condensed phase are trapped on a sorbent by pumping the headspace gas in a closed circuit via the trap and the condensed phase [20–21]. This method was initially used to determine hydrophobic compounds in water in the lower ng/l-range. A schematic representation of a closed loop strip/trap-arrangement is shown in Figure III.3.–9.

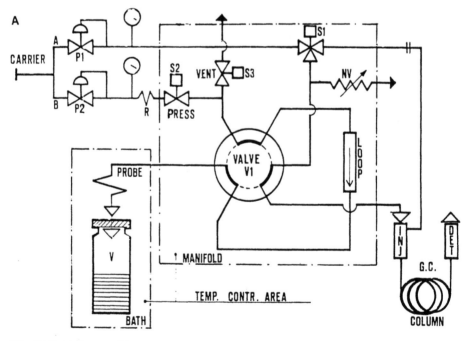

Fig. III.3.–8 Sequential purge and trap sampler-pneumatic circuit [29]

As already mentioned, adsorbed volatiles can be re-emitted into the purge gas. In "open" purge and trap arrangements, this process should be avoided because it leads to a loss of the analyte. In a closed system this can be accepted and used for analytical purposes. If the purge gas volume pumped through is large enough it is possible to bring the whole system into equilibrium, similar as described for static HSGC. Hence, the closed loop method can be performed in two ways: The experimental conditions can be optimized to collect the whole amount of volatile compounds from the sample in the trap, if possible. In this case reaching the equilibrium should be avoided by not allowing too high purge gas flow rates for example, and providing a sufficient capacity of the sorbent. According to this, an external calibration can be carried out. Particularly if polar compounds from aqueous matrices are being determined it may occur that a total extraction is not possible because comparatively long purge times are required. In such a case it is useful to optimize the system in a way that equilibrium is reached after a short purge time, e. g. by means of high gas flow rates. The calibration has to be carried out by adding standard solutions to a comparable matrix and performing the stripping-process with the standards as well as with the samples until equilibrium is reached. This system was used by Novak et al. [24] to analyze two polar compounds (ethanol, butyl acetate) from aqueous food in model experiments. Working with this system, however, is time consuming and automated devices are not available. Although this technique has found little consideration in food analysis up to now, it could offer interesting approaches for solutions to special problems: The advantages of a dynamic HS-method (good analytical sensitivity) can be linked up with those of a static method (high precision) by use of the standard addition method, especially if the capacity of the sorbent does not suffice for a complete trapping of the analytes.

3.3   Quantitation

In many cases performing quantitations according to dynamic HS-techniques, is rather awkward and take a lot more work than in automated equilibrium HSGC. With the purge and trap method, quantitations are carried out by means of external calibration. Quantitative determinations according to the standard addition method can be performed, but because of the great amount of work required are not applied in practice. This does not hold true in the case of the above described dynamic HS autosampler which allows to perform the standard addition method similar to automated equilibrium HSGC.

3.3.1   External Calibration

The instrumental aspects, discussed in III.3.2 lead to the conclusion, that before employing quantitations according to the external calibration method using "open" purge and trap-devices, certain prerequisites have to be met, for instance the stripping process has to be carefully optimized to ensure complete extraction of the volatiles that are to be determined. Parameters which can influence this process are the weight and kind of the sample (solid/liquid, particle size in suspensions), the temperature, the gas flow, intensity of stirring, and the addition of chemicals before gas extraction. Furthermore, the enrichment step has to be optimized if solid adsorbents are used, and so do the kind and quantity of the sorbent. If injection is carried out by means of thermal desorption, complete desorption must be ensured. Temperature-enhanced chemical interactions influencing analyte concentration must be avoided. If these prerequisites are fulfilled, calibration can be realized by injection of standard solutions. If a gaseous sample is injected (thermal desorption) the calibration can be carried out by injection of gaseous standards. The best way of calibration which ensures more reliable results is to analyze aqueous standard solutions in the same manner as the samples. Losses of analyte which might occur during enrichment or desorption are compensated by this calibration method. As in equilibrium HSGC the sample matrix effect can be taken into account by adding standards to an uncontaminated matrix if an exhaustive extraction is not possible. This is only useful, however, if samples of one type are analyzed. If various foods are to be analyzed, a separate calibration for each type of sample has to be accomplished, which can become very time consuming.

Generally, developing a purge and trap-procedure one would try to obtain extraction as complete as possible, check the recoveries by standard addition, and pick out the easiest external calibration method. If the nature of the problem permits it an automated equilibrium HSGC method is usually preferred. Applications of the quantitative purge and trap-analysis can be found in Chapters III.3.5.1.3 and III.3.5.5.

4 Matrix-Independent Quantitation by Means of Multiple Headspace Extraction

Multiple Headspace Extraction (MHE) is a special kind of quantitative HS analysis that allows the examination of solid samples where homogeneous partitition of added standards is impossible. The method has been developed for static HSGC and is mainly applied in this domain. The technique is based on a stepwise gas extraction of the sample with intermediate HS-analysis. Results obtained by means of external calibration do not depend on the sample matrix. Following, the performance of the MHE procedure using an electropneumatic sampling system for static HS is described [33–34]. Figure III.3.–10 illustrates the procedure according to KOLB [34].

The sample is equilibrated followed by a pneumatic HS injection. After each single injection the vessel remains under pressure because the injected aliquot is only about 10 % of the total gas volume, not large enough to cause a significant extraction yield. Therefore the vessel is vented to reach atmospheric pressure. This step is followed by reequilibra-

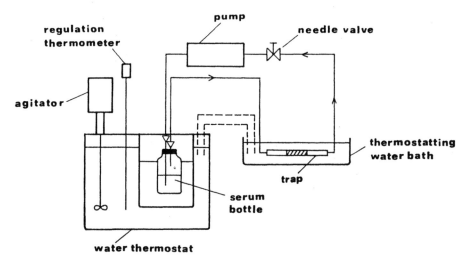

Fig. III.3-9 Schematic representation of a closed loop strip/trap-arrangement [24]

tion, repressurization, injection and reventing at defined time intervals. The external standard, in the simplest case an empty HS vessel into which a known amount of analyte has been injected, must be analyzed under the same instrumental conditions as the sample.

Assuming the gas extraction follows the mathematical description of a first-order reaction the decrease of analyte concentration in the vapor phase occurs exponentially. Therefore, sample should reach equilibrium within a not too long thermostat interval. The diffusion distances in the solid body should be as short as possible to allow a comparatively rapid transfer of the volatiles into the HS gas. For example, this is true if the sample consists of a thin film or small granules of a polymer. By extrapolation the sum of the peak area corresponding to the amount of analyte in the sample or in the standard can be calculated as the sum of a geometric progression. The peak area A_n after n extraction steps corresponds to:

$$A_n = A_1 \cdot e^{-(n-1)k} \tag{8}$$

where: A_1 = peak area of the first HS analysis, k = constant, contains partition coefficient and instrumental parameters.

If Equation (8) is developed as a geometric progression, the sum of the peak area can be calculated according to the general equation:

$$\Sigma A_n = A_1 (1-e^{-k})^{-1} \tag{9}$$

In the simplest case, the sum can be calculated after two extraction steps:

$$\Sigma A_n = \frac{A_1^2}{A_1 - A_2} \tag{10}$$

An extrapolation from two values is allowed only if the curve in the extrapolation range is known. In practice this means that for each type of sample and standard a series of extractions has to be carried out. Since the extraction procedure follows a logarithmic function, a straight line is obtained when plotting peak areas versus number of extractions on a semilogarithmic graph [4, 34]. In this case it is allowed to simplify the procedure to two extraction steps, considering the fact that extrapolation from two measurements affects the accuracy and statistical certainty. Figure III.3.–11 shows a semilogarithmic plot of the peak areas versus the number of extraction steps in the determination of styrene monomer in polystyrene. Figure III.3.–12 shows the chromatograms obtained in the determination of 12 ppb ethylene monomer in polyethylene pellets [35].

In Figure III.3.–11 a typical linear relationship can be seen. The different gradients of the sample- and the calibration graphs are characteristic of the respective matrix, while the sum of the peak area allows matrix-independent quantitation. The chromatograms shown in Figure III.3.–12 illustrate the exponential decrease of the analyte concentration in the HS gas during stepwise gas extraction. Problems which can make the MHE analysis more difficult or in some cases impossible are for example: long diffusion distances of the analyte in the solid phase, adsorption effects in the HS vessel, or changes in the matrix during the analysis [35]. If the partition coefficient is great which means that there is only

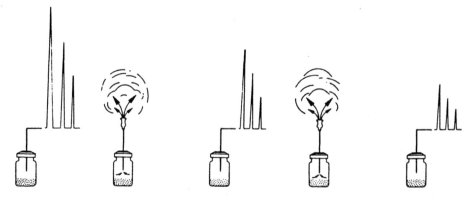

Fig. III.3.–10 Practical execution of the Multiple Headspace Extraction (MHE) procedure [34]

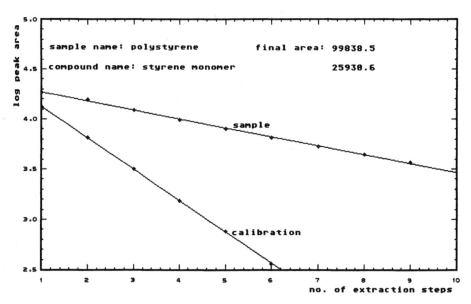

Fig. III.3.–11 Determination of 770 ppm (w/w) styrene in polystyrene by regression calculation from a nine-step MHE analysis. *Equilibration:* 100 mg, 40 min at
120 °C, *calibration external standard:* 2 µl of styrene solution in dimethyl
formamide [35]

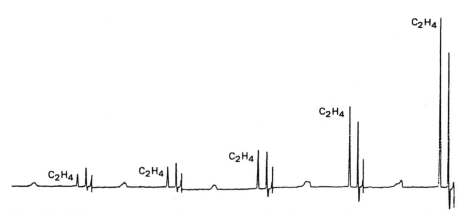

Fig. III.3.–12 Determination of 12 ppb (w/w) of ethylene in polyethylene pellets from a
five-step MHE analysis., 2 m × 1/8 in column packed with Porapak Q,
80 °C isothermal, *sample:* disks 4 mm × 2 mm, *equilibration:* 4 g, 120 min
at 120 °C [35]

a small part of the analyte in the vapor phase, the graph in the semilogarithmic plot shows a very small gradient. This results in an uncertain extrapolation so that in an extreme case the determination by MHE analysis may be impossible.

The MHE methods described above were developed using equilibrium HSGC. The previously described theoretical relationship also holds true for dynamic HS techniques in the case of incomplete extraction. A practical example for that is illustrated in Figure III.3.– 13. The first six chromatograms of a nitrogen flow passing through a solution of the four simplest aromatic hydrocarbons in squalane are shown. The concentrations of the volatiles in the vapor phase also show an exponential decrease [1].

The application of MHE to a problem of residue analysis according to a dynamic HS method is reported by Poy et al. [29]. The authors developed a method to determine VC-monomer in PVC in the lower ppb-range using a sequential purge and trap device. The MHE-procedure was performed automatically by stripping dispersions of PVC in dimethyl-acetamide (DMAC) or VC standard solutions in DMAC. This method simplified considerably the determination of VC in polymers in the lower ppb-range, because in other approaches combinations of dynamic and static HS methods have been used to improve the analytical sensitivity [36]. The application of MHE in dynamic HSGC may combine the advantages of automated MHE with those of dynamic HS techniques in which generally lower detection limits are obtained.

Finally a critical remark about the practical importance of MHE: although the method offers a very interesting approach to solving many problems it did not gain importance as a standard method in food chemistry laboratories yet, judging by the number of articles that have been published on this subject. However, this may be because the procedure is more time consuming than other methods, and the samples have to fullfil a number of prerequisites to be suitable for MHE. Though in many cases disturbing effects can be eliminated, it takes a thorough working knowledge of the method, which often is not possible for a routine working analyst. Furthermore, the fact has to be considered that only a relatively small number of analytical problems require the application of MHE. If similarly satisfactory quantitative results can be obtained using another HSGC method for reasons of saving time the latter will be preferred. Today the MHE in the field of food analysis has to be considered mainly as an interesting and important method to check and confirm other analytical methods. MHE applications are shown in the determination of solvent residues in packaging materials (c. f. III.3.5.7) and of volatile halocarbons in butter [37].

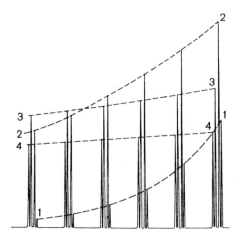

Fig. III.3.–13 Chromatograms of the vapor phase obtained on bubbling nitrogen through 3.9 ml of a solution of hydrocarbons in squalane at a rate of 55 ml/min at 30 °C. *Sampling interval:* 15 min. *Compounds:* **1** benzene, **2** toluene, **3** p-xylene, **4** cumene [1]

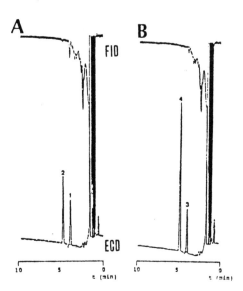

Fig. III.3.–14 HSGC analysis of wheat flour, extraction mixture acetone/water 5+1 (v+v), **A** *Standard:* **1** bromotrichloromethane (internal standard), **2** tetrachloroethene 195 µg/kg, **B** *Sample:* **3** bromotrichloromethane (internal standard), **4** tetrachloroethene 458 µg/kg [45]

5   Applications

5.1   Volatile Halocarbons

Volatile halocarbons (VHCs) such as trichloromethane (chloroform), tetrachloromethane (carbon tetrachloride), 1,1,1,-trichloroethane, trichloroethene, and tetrachloroethene (perchloroethylene) are primarily used as industrial solvents and chemical intermediates. More than 10 million tons of these compounds are produced annually all over the world. In addition, trihalomethanes may be produced through the chlorination of water. VHCs are of interest, because their large production volumes may lead to their presence as contaminants in foods. Some have been found to be carcinogenic in animals. Some special VHCs 1,2-dibromoethane (ethylene dibromide), 1,2-dichloroethane (ethylene dichloride), bromomethane (methyl bromide) or tetrachloromethane are used as fumigants against insect pests in stored food commodities like grains or fruits. This has led to the development of a number of different HS methods to determine VHCs in food [38−41]. In addition, some of these compounds have found technological use in the food industry which made it necessary to develop methods for the determination of e. g. dichloromethane (methylene chloride) in hop extract [42] or of trichloroethene or dichloromethane in decaffeinated coffee [43]. Tetrachloroethene (Perchloroethylene) is extensively used as a solvent in dry cleaning. Foods, that are stored or sold within the emission area of dry cleaning units may therefore be contaminated with this solvent [44].

5.1.1   Determination of Tetrachloroethene (TCE) in Wheat Flour Using the Method of External Calibration

Volatile halocarbons can be easily extracted from grains or flour using a mixture of acetone and water (5:1, v/v) [46]. For example, 2 g of flour sample are weighed into a 20 ml HS vessel and 3 ml of a solution of bromotrichloromethane (BTCM, internal standard) in acetone and 0.6 ml water are added, and the vessel is sealed. For calibration known amounts of TCE in acetone (which also contains the internal standard in the same concentration as the sample) and water are added to 2 g of uncontaminated wheat flour. The standard addition should correspond to 20−500 µg TCE/kg in the samples. Standards and samples are treated in an ultrasonic bath for 15 min and then stored for at least 24 h. For determination an equilibrium HS autosampler is used. The vessels are kept at 60 °C for 90 min and splitless injection is performed. The separation is carried out on a 25 m × 0.32 mm CP-Sil 8 fused silica capillary column with a film thickness of 5 µm at a temperature of 120 °C (isothermal). Helium can be used as carrier gas. For detection an ECD at a temperature of 300 °C is used [45]. Figure III.3.−14 shows the gas chromatograms of a standard and a TCE-contaminated sample. Figure III.3.−15 shows a calibration graph obtained.

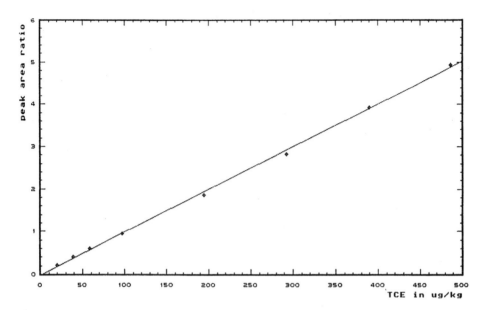

Fig. III.3.–15 Determination of tetrachloroethene in wheat flour, calibration graph, peak area ratio plotted versus concentration in μg/kg [45]

Other application examples for the determination of VHCs in foods using external calibration are listed in Table III.3.–1.

5.1.2 Analysis of VHCs in Various Foods Using the Standard Addition Method

According to the method described in III.3.5.1.1, the quantitation can also be carried out by the standard addition method by simply adding known amounts of standard to the sample itself. A calibration graph, obtained in the determination of TCE in wheat flour by means of the standard addition method is shown in Figure III.3.–16.

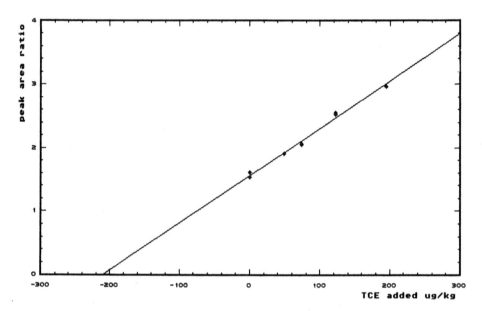

Fig. III.3.–16 Determination of tetrachloroethene in wheat flour using the method of standard addition, peak area ratio plotted versus concentration, sample containing 210 µg/kg TCE [45]

A method to determine VHCs in various foods was developed by ENTZ and HOLLIFIELD [38]:

Solids such as meat and fish are ground while partially frozen. Specific food samples are treated in different fashion depending on the physical nature of the sample. Liquids are analyzed undiluted. Semisolids (e. g. butter) or viscous liquids, if free flowing at 90 °C are analyzed undiluted also. Water miscible foods (e. g. jellies) are diluted with distilled water or, if waterimmiscible (e.g. meat) digested in 20 N H_2SO_4. The general procedure is as follows: 1–2 g of the sample is weighed into a tared crimp-top vial; 1–5 ml of distilled water or 15 ml of digestion medium is added to the vial if solvation or digestion is needed. Internal standard (1,1,1,3-tetrachloro-tetrafluoropropane) and spiking standard solutions in 2-propanol are added with microliter syringes. The vials are sealed with crimp seals containing teflon-lined septa, shaken, and equilibrated at 90 °C for 1 h. 2 ml of headspace gas are injected manually (packed columns: 15 % OV-17 on 80–100 mesh Chromosorb W HP or 20 % SP-2000 / 0.1 % Carbowax 1 500 on 100–120 mesh Supelcoport). The authors have also used an automated equilibrium headspace analyzer which was equipped with a OV-101 glass capillary column (0.27 mm i. d., film thickness approx. 0.9 μm). In this case, injection was performed at an initial pressure of 2.1 bar and a split ratio of 3.5 : 1. Injection time was 8 seconds. Separation can be carried out using the described capillary column by injecting at 80 °C oven temperature, and heating at a rate of 6 °C/min to 116 °C and holding at this temperature for another 2 min. Argon/methane (5 %) was used as carrier- and make up gas. Figure III.3.–17 illustrates the analysis of trichloro-ethene (TRI) in a crab apple jelly sample using this method. The concentration of the TRI residue was determined to be 26 μg/kg.

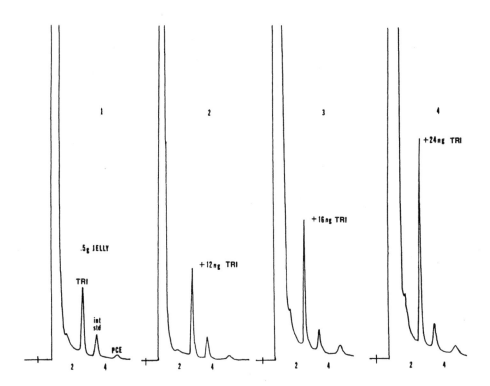

Fig. III.3.–17 Analysis of crab apple jelly using the standard addition method (column: 15 % OV–17 3.7, m 90 °C), **1** 0.5 g of jelly in 5 ml of water, **2** 12 ng of TRI added, **3** 16 ng of TRI added, **4** 24 ng of TRI added [38]

Table III.3.–1 Determination of VHCs in Foods Using Static Headspace Methods

Subject	Quanti-tation	Remarks	Authors
Volatile halocarbons in various foods	1,2	autom./man inj.	Köbler et al., 1982 [17]
Volatile halocarbons in various foods	1,2	autom./man. inj.	Entz and Hollifield, 1982 [38]
Volatile halocarbons in various food composites	1,2	autom./man. inj.	Entz. et al., 1982 [18]
Volatile halocarbons in butter	2,3	autom. inj.	Uhler and Miller, 1988 [37]
Methyl bromide in food ingredients	1	autom./man. inj.	de Vries et al., 1985 [39]
Methyl bromide in grapefruit	1	man. inj.	King et al., 1981 [40]
Ethylene dibromide in cereals	1	autom. inj.	Pranoto-Soetardhi et al., 1986 [41]
Ethylene dibromide in fruits	1	autom. inj.	Gilbert et al., 1985 [47]
Methylene chloride in hop extract	1	man. inj.	Neumann and Wagner 1974 [42]
1,2-Dichloroethylene in instant coffee	3	autom. inj.	Kolb et al., 1984 [35]
Dichlorodifluoromethane (freezant) in contact-frozen prawns	1	man. inj.	Carter and Kirk, 1977 [16]

1: External Calibration, 2: Standard Addition Method, 3: Multiple Headspace Extraction

5.1.3 Determination of VHCs in Foods Using Purge and Trap Procedures

The "classical" application of purge and trap methods in residue analysis is the determination of (mainly non-polar) volatile organic contaminants in water. For this mostly organic polymers are used as sorbents and injections are carried out by thermal desorption. Standard methods of the US-Environmental Protection Agency (EPA) are based on this technique [48]. Contaminated water represents an ideal matrix for the analysis of VHCs: non-polar compounds can effectively be extracted from aqueous solutions and trapped by organic sorbents so that outstanding results can be obtained. In the beginning of the 80's purge and trap methods were employed for the determination of VHC residues in food for the first time. In two extensive research programs, carried out simultaneously to estimate the health and ecological risks caused by these compounds a great number of foods were examined for their background pollution with VHC [30, 49–54].

A method developed by Bauer [50,52] allows the determination of up to 33 halogenated and nitro compounds within one GC separation, and this method is described in the following.

10 g of a frozen solid sample are ground with pre-cooled sodium sulfate using a mortar with pestle. During this procedure the sample temperature should not rise above 4 °C. The sample is stripped for 40 min with 40–50 ml N_2/min in a 100 ml glass flask equipped with a fritted bottom and maintained at 80 °C by means of a water bath. The volatiles are adsorbed in a XAD-2-tube at 0 °C and eluted with 10 ml of pentane. An aliquot is analyzed by GC (60 m × 0.25 mm OV-101 capillary column; carrier gas helium, 4 ml/min; oven temperature: 10 min 85 °C, 5 °C/min to 140 °C, 9 min isothermal). Figure III.3-.18 shows the design of the purge and trap apparatus used. The separation of 33 halogenated hydrocarbons and nitro-compounds is illustrated by the chromatogram in Figure III.3.–19. For the various foods recoveries are checked by adding pentane standard solutions to the

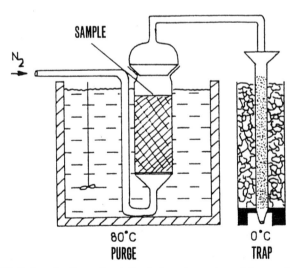

Fig. III.3.–18 Determination of volatile halocarbons in solid samples, purge and trap apparatus [25]

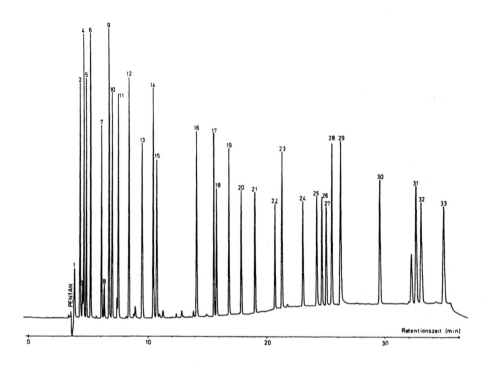

Fig. III.3.–19 Determination of volatile contaminants in food using a purge and trap procedure, gas chromatographic separation of 33 compounds [50]

List of compounds.: **1** dichloromethane, **2** trichloromethane, **3** 1,2-dichloroethane, **4** 1,1,1-trichloroethane, **5** tetrachloromethane, **6** trichloroethene, **7** 1,1,2-trichloroethane, **8** 1,3-dichloropropane, **9** bromodichloromethane, **10** 1,2-dibromoethane, **11** tetrachloroethene, **12** 1,1,1,2-tetrachloroethane, **13** tribromomethane, **14** 1,1,2,2-tetrachloroethane, **15** 1,2,3-trichloropropane, **16** pentachloroethane, **17** 1,3-dichlorobenzene, **18** 1,4-dichlorobenzene, **19** 1,2-dichlorobenzene, **20** tetrabromomethane, **21** hexachloroethane, **22** 2,4 dichlorotoluene, **23** 1,3,5-trichlorobenzene, **24** 1,2,4-trichlorobenzene, **25** 4-nitrotoluene, **26** 1,2,3-trichlorobenzene, **27** 4-chloronitrobenzene, **28** hexachlorobutadiene, **29** 1,3-dichloro-3-nitrotoluene, **30** 5-chloro-3-nitrotoluene, **31** 4-chloro-3-nitrotoluene, **32** 1,2-dichloro-3-nitrobenzene, **33** 2-chloro-4-nitrotoluene sample before stripping. Detection limit depends on the compound to be determined, i. e. Tetrachloromethane: 0.1 µg/kg, 1,2-dibromoethane: 0.8 µg/kg.

Some applications of purge and trap-procedures in the determination of VHCs in various foods are listed in Table III.3.–2.

Table III.3.–2 Determination of Volatile Halocarbons in Foods Using Purge and Trap Methods

Subject	Remarks	Authors
Volatile halocarbons in various foods, water and air	self-designed apparatus, purge gas N_2, sample temp. 80 °C, sorbent XAD 2, pentane elution.	Bauer, 1981 [50, 52]
Volatile halocarbons in human and animal tissues	self-designed apparatus, purge gas N_2, sample temp. 80 °C, sorbent XAD 2, pentane elution.	Alles et al., 1981 [25]
Volatile halocarbons in various foods, water and air	self-designed apparatus, purge gas N_2, sample temp > 90 °C, vig. stirring, add. of proteolyt. enzyme before outgasing, add. of detergent and acid, sorbent XAD 2, pentane elution.	Lahl et al., 1981 [30]
Ethylene dibromide in table-ready foods	self-designed apparatus, purge gas N_2, water bath 100 °C, stirring, sorbent Tenax TA, hexane elution ECD, HECD, GC MS	Heikes, 1985 [55]
Ethylene dibromide in grains and animal feeds	self-designed apparatus purge gas N_2, water bath 100 °C, stirring, sorbent Tenax TA, hexane elution.	Heikes, 1985 [56]
Fumigants in grains	cf. [56] but using Tenax TA/XAD 4 comp. as sorbent	Heikes and Hopper, 1986 [31]
Chlorinated solvents in decaffeinated coffee	cf. [31]	Heikes, 1987 [43]
Volatile halocarbons and CS_2 in table-ready foods	cf. [31]	Heikes, 1987 [57]
39 volatile org. pollutants in fish	commercially available apparatus purge gas N_2, trap Tenax GC/silica gel/ act. charcoal, thermal desorption GC-MS	Easley et al., 1981 [56]

5.2 Solvent Residues in Oils and Fats

Some examples for the determination of residues of extraction solvents are given in Table III.3.-3. In all cases, static HS methods were used.

Table III.3.–3 Solvent Residues in Oils and Fats

Subject	Quanti-tation	Remarks	Authors
Solvent residues in oils and fats	1	autom. inj.	Nosti Vega et al., 1970 [59]
Solvent residues in oils and fats	1	autom. inj.	Belluco et al., 1979 [60]
Solvent residues in oilseed meals and flours	1	man. inj.	Dupuy and Fore, 1970 [14]

1: External Calibration

5.3 Polar Compounds in Aqueous Foods

Equlibrium HSGC was originally developed for the determination of ethanol in blood [61–62] and today it is the standard method throughout most of Europe for this application. Practical applications of HS analyses of polar compounds from aqueous matrices are the determination of methanol in wine by automated equilibrium HSGC [63] or the quantitation of ethanol as a chemical index for decomposition in canned salmon [64]. In addition, the determination of carbonyl compounds such as acetaldehyde in aqueous foods by means of static HSGC has been carried out [65]. As an example, the determination of methanol in wine is described in the following:

To 1 ml of wine in a 6 ml HS-vessel 0.3 ml water, 0.2 ml of an aqueous solution of t-butanol (internal standard) and, in order to destroy the interfering acetaldehyde peak, 0.2 ml 0.1 ml silver nitrate solution and 10 µl of aqueous sodium hydroxide (32 %) are added. The standard samples are obtained by adding all reagents as well as known amounts of methanol to an uncontaminated wine (standard concentration: 200–1 000 mg/l). Samples and standards are sealed immediately after preparation, equilibrated at 80 °C for 20 min and the injection (split mode) is carried out by means of an electropneumatic HS-autosampler. For separation, an OV-101 fused silica capillary (I. D.: 0.23 mm, film thickness 0.4 µm) is used. The HS gas chromatograms of one silver nitrate treated and one untreated wine sample are shown in Figure III.3.–20 [63].

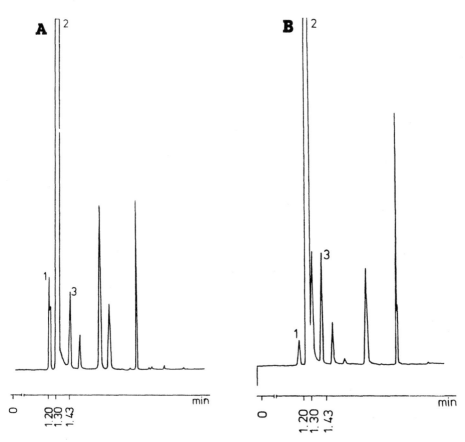

Fig. III.3.–20 Headspace gas chromatographic analysis of wine samples, **A** untreated: **1** methanol/acetaldehyde, **2** ethanol, **3** t-butanol (internal standard); **B** after AgNO₃-treatment: **1** methanol 150 mg/l, **2** ethanol 9.2 vol %, **3** t-butanol (internal standard) [63]

5.4   Dithiocarbamate Fungicides in Foods

Metal salts of the alkylene-bis-dithiocarbamic acids have been used as fungicides since about 1930. They are mainly used in fruit- and vegetable cultivation. Analysis is generally carried out by determining the decomposition product carbon disulfide (CS_2). For the determination of CS_2 a colorimetric method can be used [66] as well as equilibrium HSGC using external calibration. Before HS analysis the sample is usually treated with HCl/SnCl₂ or H_2SO_4/SnCl₂ [13, 66–68] to volatilize the CS_2. A flame photometric detector in the sulphur mode is suitable for detection. BLAICHER et al. [67] have used a potassium sulphate doted flame ionization detector. Their method is described here:

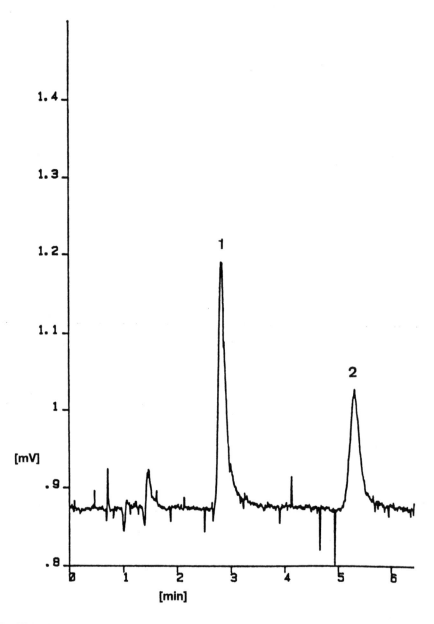

Fig. III.3.–21 Determination of dithiocarbamates in foods: Chromatogram of a standard, **1** CS_2, concentration is equivalent to 1.26 mg/kg; **2** thiophene (internal standard) [67]

If possible foods should be analyzed just after delivery. 10 g of food are weighed into a 100 ml flask. Leafy vegetables are torn into little pieces and, in case of grapes the entire fruit is used. In case of apples only the peal is used, however, the results are calculated for the whole fruit. To decompose the dithiocarbamates, 50 ml of a 1.5 g $SnCl_2$ in 100 ml 4 N hydrochloric acid solution are added. The flask is closed by a screw cap equipped with a septum (pierced stopper). As internal standard 10 µl of a solution consisting of 20 µl thiophene in 10 ml methanol are injected through the septum with a microliter syringe. Calibration standards are prepared by adding known amounts of carbon disulfide in methanol (corresponding to 0.63–2.52 mg/kg in 10 g sample) as well as the internal standard into closed sample flasks which already contain the reaction mixture by means of a microliter syringe. Samples and standards are equilibrated at 80 °C for 2 hours and 0.5 ml HS gas are injected into the GC using a gas tight syringe. The separation is carried out on a 3.8 m steel column (⅛") on 30 % DC 200 on Chromosorb W/AW-DCMS at an oven temperature of 80 °C, using nitrogen as carrier gas. The GC separation is shown in Figure III.3.–21.

5.5   N-Nitrosamines in Animal Feeds

N-nitroso compounds have shown to be carcinogenic in various animal species. N-nitrosamines may be formed by the reaction of primary, secondary and tertiary amines with various nitrosating agents e. g. nitrous acid, nitrogen oxides, and the nitrous acidium ion. Nitrites and/or nitrates have been used as preservatives or color fixatives in fish and meat products. From these additives nitrous acid can arise, which then may react with amines in the products to form N-nitrosamines. This has prompted interest in their determination in foodstuffs [69] and animal feeds, especially if they contain nitrate-treated fish meal added as a protein source. N-nitrosamines in animal feeds are also a serious problem for researchers maintaining laboratory animals for use in chemical carcinogenesis research. BILLEDEAU et al. [70] have developed a high temperature purge and trap procedure for determining seven volatile nitrosamines in animal feed, using a gas chromatograph equipped with a thermal energy analyzer (TEA):

A 5 g animal feed sample is accurately weighed into a 50 ml graduated impinger tube. 0.5 g sulfamic acid is added as nitrosation inhibitor. 20 ml mineral oil is added, and the contents are thoroughly mixed. Samples are purged with 400 ml/min Ar at 150 °C for 1 h. The N-nitrosamines (N-dimethylnitrosamine NDMA, N-diethylnitrosamine NDEA, N-dipropylnitrosamine NDPA, N-dibutylnitrosamine NDBA, N-nitrosopiperidine NPIP, N-nitrosopyrolidine NPYR, N-nitrosomorpholine NMOR) are collected on Thermo-Sorb/N-cartridges. After purging, the cartridges are eluted by back flushing with acetone-dichloromethane (1:1, v/v). 8 µl of this solution are injected in the GC/TEA-System. Calibration is performed with external standards. The authors have used 10 % Carbowax 20M - 2 % KOH on 80–100 mesh Chromosorb W (AW) as stationary phase. The carrier gas was Ar at a flow of 40 ml/min. The column was operated in a temperature range from 150 °C to 190 °C at 4 °C/min. The TEA pyrolysis chamber was kept at 500 °C. A diagram of the impinger-tube is shown in Figure III.3.–22. Figure III.3.–23 shows the structures of the seven volatile nitrosamines. The GC/TEA chromatograms of a standard mixture, one spiked and one unspiked feed extract are presented in Fig. III.3.–24.

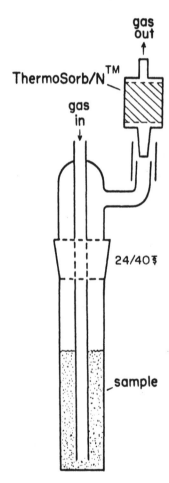

Fig. III.3.–22 Diagram of closeup of impinger tube fitted with ThermoSorb/N cartridge [70]

H₃C
  \
   N–N=O
  /
H₃C

NDMA

H₅C₂
  \
   N–N=O
  /
H₅C₂

NDEA

N–N=O

NPIP

N–N=O

NPYR

H₇C₃
  \
   N–N=O
  /
H₇C₃

NDPA

H₉C₄
  \
   N–N=O
  /
H₉C₄

NDBA

O   N–N=O

NMOR

Fig. III.3.–23 Structures of seven volatile N-nitrosamines [70]

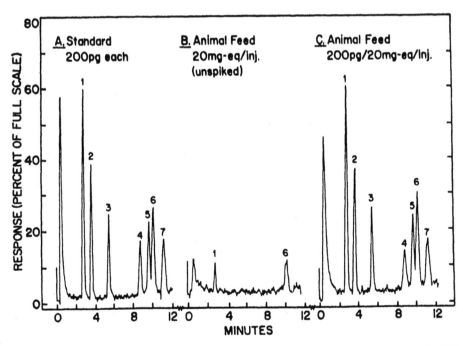

Fig. III.3.–24 GC/TEA chromatograms of **A** standard mixture, response represents 200 pg each of **1** NDMA, **2** NDEA, **3** NDPA, **4** NDBA, **5** NPIP, **6** NPYR, **7** NMOR; **B** feed extract; **C** spiked feed extract [70]

5.6   Methylmercury in Biological Samples

Methylmercury compounds belong to the most dangerous substances in the environment. They are highly toxic, even more than inorganic mercury, and are often found to be concentrated at the end of the food chain. A semiautomated equilibration HSGC procedure for the determination of methylmercury in biological samples using a microwaveinduced plasma detector system was developed by DECADT et al. [71]:

A 0.5 ml portion of homogenated aqueous sample is added to 0.5 ml water or methylmercury standard solution (standard addition method) in a HS vessel. Addition of 18.5 mg of iodoacetic acid and 12.5 mg sodium thiosulfate (prevents photochemical degradation of MeHgI) results in the liberation of methylmercury iodide. The vessel is closed immediately and held for 5 min at 70 °C. A semiautomated headspace sampler is used. Sufficient sample quantity is delivererd after an injection time of 15 seconds and a pressure of 1.3×10^5 Pa in the HS vessel. GC analysis can be carried out using a 1 m

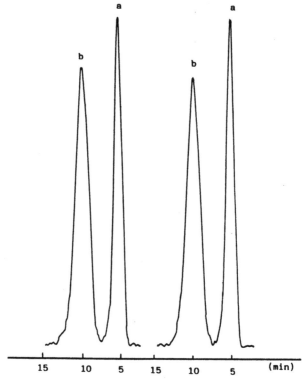

Fig. III.3.–25 Gas chromatogram of methylmercury iodide **(a)** and ethylmercury iodide **(b)** [71]

(3 mm i.d.) Teflon column packed with 10 % AT-1 000 on Chromosorb WAW 80/100 mesh and 100 ml/min argon as carrier gas. Emmission measurements are carried out with a Perkin Elmer AAS-403 at 253.7 nm. Typical chromatograms are shown in Figure III.3.–25. The best results are obtained when performing calibration according to the standard addition method.

5.7 Residual Solvents in Packaging Materials

Food packaging materials often consist of thin aluminium or plastic films which carry printing on the outside. The packaging material itself as well as the printing ink or the

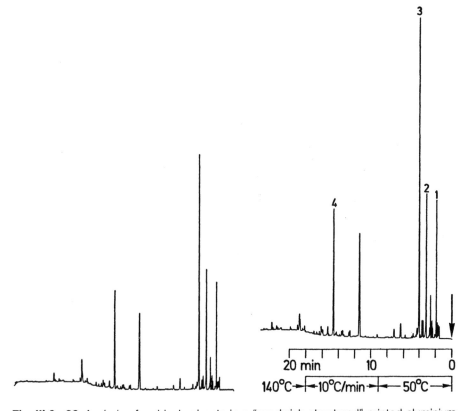

Fig. III.3.–26 Analysis of residual solvents in a "sandwich structured" printed aluminium film by multiple headspace extraction; the first two chromatograms from a five-step analysis. Column temperature programmed as indicated, for conditions see text; *sample:* 50 cm², *Compounds:* **1** n-hexane 1.8 mg/m², **2** methanol 9 mg/m², **3** ethanol 12 mg/m². **4** ethyl glycol 6 mg/m² [72]

adhesive (in laminated materials) may contain residual solvents, which can migrate into the foods. In many cases with such materials no calibration can be performed and often it is impossible to dissolve the entire sample before HS analysis. Since packaging materials are mostly thin films, MHE can be performed very well. KOLB et al. [72] have used the MHE technique to determine residual solvents in printed aluminium films:

50 cm^2 "sandwich-structured" printed aluminium film are equilibrated in a closed HS vessel at 140 °C for 45 min. The MHE procedure is performed as described in Chapter III.3.4. For calibration some microliters of a standard mixture are injected in an empty HS vessel and analyzed in the same way as the sample. The separation of hexane, methanol, ethanol and ethyl glycol can be carried out on a 35 m × 0.3 mm glass capillary with Marlophen M 87 stationary phase (for temperatures see chromatogram). For quantitation, the total area is calculated by means of the extrapolation method described in III.3.–4 (calculating the peak area which corresponds to total extraction of the analytes). Figure III.3.–26 shows the HS gas chromatograms of the first two extraction steps from a five step MHE analysis.

The authors also describe the analysis of ethyl acetate in printed aluminium films. The determination of residual solvents from thin films by means of MHE is reported in further papers [73–74].

5.8   Monomer Residues in Polymers and Foods

The determination of monomer residues in polymers and foods is another important domain of HSGC. Because many monomers are toxic or may cause cancer (e. g. styrene, vinyl chloride), a rapid and precise control of residues is particularly important. In Table III.3.–4 applications of HS methods in this field are compiled. Polymers are often dissolved or finely dispersed in organic solvents prior to analysis.

As an example for the analysis of monomer residues, the determination of acrylonitrile in an acrylonitrile-butadiene-styrene (ABS)-resin and in olive oil is described according to the method of DI PASQUALE et al. [78]:

To 5 ml dimethylsulphoxide (DMSO) in a 10 ml HS vessel 0.5 g ABS-resin is added and the vessel is sealed. When a fine dispersion is obtained, 10 µl of a solution of propionitrile in DMSO (internal standard) are injected into the liquid. Solutions of acrylonitrile in DMSO are used as standards to which the internal standard is added in the same manner. The vessels are stirred for 5 min and equilibrated at 80 °C for 60 min. The automated injection was followed by a separation on a 2 m × 2 mm I. D. steel column containing 10 % Carbowax 1 500 on Chromosorb WAW (60–80 mesh). As detectors, the FID or the nitrogenselective detector (NSPD) for improved detection limit are used. If the FID detection is employed, dimethylformamide with 10 % water is used as solvent. Figure III 3.–27 shows a calibration curve obtained with this method. To determine acrylonitrile in olive oil, 5 ml of sample are held for 60 min at 80 °C and subsequently analyzed in the same manner. Non-contaminated olive oil is used to prepare calibration standards.

Table III.3.–4 Determination of Monomer Residues in Polymers and Foods
A: Static HS-Methods

Subject	Quanti-tation	Remarks	Authors
Vinyl chloride in foods	2	man./autom. inj., collaborative study	Biltcliffe and Woods, 1982 [75]
Vinyl chloride in PVC	1	autom. inj.	Krockenberger and Gmerck, 1987 [76]
Vinyl chloride and acrylo-nitrile monomer in foods and polymers	1	autom. inj.	Chudy and Crosby, 1977 [77]
Acrylonitrile monomer in ABS resins and olive oil	1	autom. inj.	Di Pasquale et al., 1978 [78]
Methyl methacrylate, toluene and styrene from plastic containers in maple syrup	1,2	man. inj.	Hollifield et al., 1980 [79]
Butadiene monomer in plastics and food	1	autom. inj., MS single ion monitoring	Startin and Gilbert, 1984 [80]
Butadiene monomer in in plastics and food	1,2	man. inj.	McNeal and Breder, 1987 [81]
Styrene monomer in foodstuffs	1	man. inj.	Santa Maria et al., 1986 [82]
Styrene in polystyrene ethylene in polyethylene	3	autom. inj.	Kolb et al., 1984 [35]
Migration of benzene from styrene into foodstuffs	1	man. inj.	Varner and Breder, 1984 [83]
Ethylene oxide in food products in heat sealed packages	1	man. inj.	Ricottilli et al., 1981 [84]

1: External Calibration, 2: Standard Addition Method, 3: Mulitple Headspace Extraction

B: Dynamic HS Methods

Subject	Remarks	Authors
≥ 1 ppb vinyl chloride in PVC food packaging	packaging materials dissolved in DMAC, He-sparged VC collected in ethanol, sealed vials are used for static HSGC	Dennison et al., 1978 [36]
Vinyl chloride in PVC at low ppb-levels	PVC dissolved in DMAC, sequential purge and trap autosampler, MHE	Poy et al., 1987 [29]

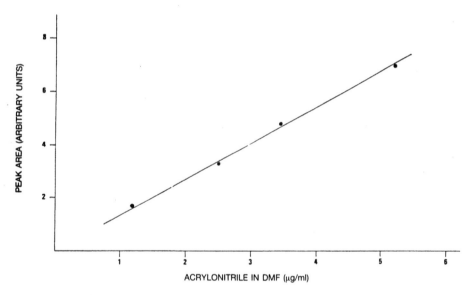

Fig. III.3.–27 Calibration curve of peak area versus the amount of acrylonitrile in the standard solution (FID-detection) [78]

5.9 Ethylene Oxide in Surfactants and Demulsifiers

The analysis of cleansing agents and cosmetics is also a task of a food chemist. The examination of ethoxylated surfactants and demulsifiers for residual ethylene oxide, which is highly toxic and causes cancer in laboratory animals may serve as an example from this field. The method was developed by DAHLGRAN and SHINGLETON [85] including a static HSGC method with external calibration or standard addition.

Calibration standards are prepared as follows: a sample of surfactant material is vacuum-stripped at 50–60 °C on a rotary evaporator to remove any residual ethylene oxide. 2 g of this material are weighed into headspace vessels, the vessels are sealed with teflon-lined silicone septa and known amounts of ethylene oxide in 10 µl hexane are injected.

To prepare the samples, 2g of surfactant are weighed into a vial. 10 µl of dried hexane are injected through the septum to ensure that standards and samples are of identical matrix. The authors have used an electropneumatic HS autosampler. Samples are equilibrated at 100 °C for 30 min. After a pressurization period of 0.7 min, injection time is 0.08 min. For separation a packed column with 80–100 mesh Chromosorb 102 at a carrier glass flow of 30 ml/min can be used. The oven temperature is kept at 120 °C for 5 min, heated to 190 °C at a rate of 8 °C/min and held for 10 min. Under these conditions, the retention time of ethylene oxide is 6–7 min. The detection limit is about 1 mg/kg (FID-detection).

The main problem is to prepare a solution with a known amount of ethylene oxide. The standard preparation method developed by the authors is illustrated in Figure III.3.–28.

An ethylene oxide cylinder is connected to a septum-sealed 10 ml hypovial using a short piece of Tygon tubing. A hypodermic needle is inserted nearly to the bottom of the vial. A second piece of tubing connects the top of the vial and a beaker of water (Figure III.3.–28 A). The apparatus is slowly purged for 15 min to remove air. At this point, a dry needle is attached to the end of the second piece of tubing, inserted into a sealed and tared vial containing 10 ml of dried hexane and the flow is increased. 40–80 mg of ethylene oxide are added to the headspace above hexane (Figure III.3.–28 B). The vial cooled to 0°C and placed 5 min on a shaker. Additional standards are prepared by diluting this primary standard.

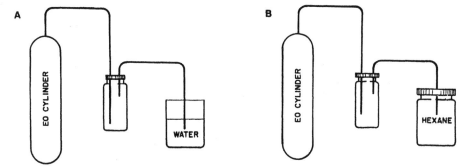

Fig. III.3.–28 Preparation of an ethylene oxide standard solution **A** Purging gas lines to remove air from tubing; **B** Addition of ethylene oxide to dry hexane [85]

6 Discussion

The equilibrium HSGC represents a well established and elegant method to determine residues of volatile compounds in various foods and packaging materials. Contamination of the analytical column by trace or non-volatile constituents of the sample are avoided by injecting inert gas instead of a solvent. Generally, very reliable results are obtained with this technique and in many cases the reproducibility is higher than in other GC methods. The static HS method is mainly suitable for sequential analysis of a great number of samples of the same type because calibration depends on the sample matrix, and in this special case can be used for the whole series of experiments. Moreover, static HS is easy to automize. On the other hand, if one compound is to be determined in many different sample matrices, work is getting much more time consuming.

The Multiple Headspace Extraction (MHE) allows a matrix-independent quantitation by means of external calibration. The method has to be carefully optimized for each sample type which first of all requires a series of analyses of at least one sample and one standard. The great advantage of MHE is that it also allows quantitative determinations of samples, on which the common calibration methods of HSGC cannot be applied.

Especially in cases of insufficient detection limits of static HSGC methods, dynamic HS techniques can be applied with reliable results. In the field of residue analysis in foods and packaging materials mainly purge and trap methods are used. Under certain conditions such methods allow the total extraction of volatile compounds from suspensions of solid samples as well. Then quantitation can be performed without considering the matrix which represents an advantage compared to static methods. The necessity of an enrichment step and the fact that automatized sequential analyses are not applicable with most of the equipment make the purge and trap methods more time and work consuming than equilibrium HSGC techniques. Furthermore, the enrichment step has to be carefully optimized, especially if organic polymers are used as sorbents. The analysis of polar volatile compounds causes problems because on one hand they are difficult to strip if the matrix contains water (large purge gas volumes are necessary) and on the other hand it is relatively difficult to trap these compounds, also if cryofocusing is applied [23].

A comparision between static HSGC and an on line purge and trap method revealed that the GC profiles of the two methods differ [32]. In comparison to equilibrium HSGC, the highly volatile compounds are more abundant and the less volatile compounds are less abundant with the purge and cold trap method. The absolute amount of a volatile compound analyzed by purge and trap/thermal desorption GC within one GC separation is considerably higher (if no split is used) than in static HSGC methods. Under favorable conditions the detection limit in dynamic HS techniques can be up to two orders of magnitude lower compared to a static method without cryofocusing. In the analysis of volatile compounds dynamic HSGC in combination with thermal desorption today represents the most sensitive technique.

First attempts to perform automatized sequential analysis in dynamic HS methods have not yet been established in the routine residue analysis. Today automatized equilibrium HSGC is the method of choice if the analytical problem can be solved by this technique.

References

1. IOFFE, B. V. (1984) J. Chromatogr. **290**: 363

2. WITTKOWSKI, R. (1987) in "Schnellmethoden zur Beurteilung von Lebensmitteln und ihren Inhaltsstoffen" (Baltes, W., Ed.), Verlag Behr's, Hamburg, p. 261

3. HACHENBERG, H., SCHMIDT, A. P., (1977) "Gas Chromatographic Headspace Analysis", Wiley Heyden, Chichester, New York, Brisbane, Toronto, Singapore

4. KOLB, B. (1986) in "Gaschromatographie mit Kapillartrennsäulen" (Günther, W., Schlegelmilch, W., F., Eds.), Vogel Würzburg, p. 149

5. KOLB, B., LIEBHARDT, B., ETTRE, L. S. (1986) Chromatographia **21**: 305

6. TAKEOKA, G., JENNINGS, W. G. (1984) J. Chrom. Sci. **22**: 177

7. BADINGS, H. T., DE JONG, C., DOOPER, R. P. M. (1985) J. High Res. Chromatogr. **8**: 755

8. ETTRE, L. S., PURCELL, J. E., WIDOMSKI, J., KOLB, B. POSPISIL, P. (1980) J. Chromatogr. Sci. **18**: 116

9. COBELLI, L., BANFI, S., (1987) International Laboratory **9**: 46

10. HUBER, L. (1984) Labor Praxis **3**: 152

11. VITENBERG, A. G., REZNIK, T. L. (1984) J. Chromatogr. **287**: 15

12. CHIALVA, F., DOGLIA, G. GABRI, G., ULIAN, F. (1983) Proc. 5th Int. Symp. on Capillary Chromatogr., Elsevier, Amsterdam, p. 430

13. McLEOD, H. A., McCULLY, K. A. (1969) J. Assoc. Off. Anal. Chem. **52**: 1226

14. DUPUY, H. P., FORE, S. P. (1970) J. Amer. Oil Chem. Soc. **47**: 231

15. BROCKWELL, C. A. (1978) J. Agric. Food Chem. **26**: 962

16. CARTER, J. R., KIRK, R. S. (1977) J. Food Technol. **12**: 49

17. KÖBLER, H., HEFFTER, A., KRITZLER, A. (1982) "Bestimmung von niedrigsiedenden Chlorkohlenwasserstoffen in fetthaltigen Lebensmitteln", Bericht zum Forschungsvorhaben BGA C-Verw. 2-1048-07-3371/82, Chemische Landesuntersuchungsanstalt Stuttgart

18. ENTZ, R. C., THOMAS, K. W., DIACHENKO, G. W., (1982) J. Agric. Food Chem. **30**: 846

19. BELLAR, T. A., LICHTENBERG, J. J., KRONER, R. C. (1974) J. Amer. Water Works Assoc. **66**: 703

20. GROB, K. (1973) J. Chromatogr. **84**: 255

21. GROB, K., Grob. G. (1975) J. Chromatogr. **106**: 299

22. DROZD, J., NOVAK, J. (1979) J. Chromatogr. **165**: 141

23. NOIJ, T., VAN ES, A., CRAMERS, C., DOOPER, R., RIJKS, J., (1986) Proc. 7th Int. Symp. Capillary Chromatogr., The University of Nagoya Press, p. 64

24. NOVAK, J., GOLIAS, J., DROZD, J. (1981) J. Chromatogr. **204**: 421

25. ALLES, G., BAUER, U. SELENKA, F. (1981) Zbl. Bakt. Hyg. I Abt. Orig. B. **174**: 238

26. Supelco Reporter International (1987) Vol VI, **4**: 8

27. BADINGS, H. T., DE JONG, C., DOOPER, R. P. M., DE NIJS, R. C. M. (1985) in: Proceedings of the 4th Weurman Flavour Research Symposium (Adda J., Ed.), Dourdan, France, Elsevier, Amsterdam, p. 523

28. POY, F., COBELLI, L. (1985) J. Chromatogr. Sci. **23**: 114

29. POY, F., COBELLI, L., BANFI, S. FOSSATI, F. (1987) J. Chromatogr. **395**: 281

30. LAHL, U., BÄTJER, K. V. DÜSZELN, J., GABEL, B., STACHEL, B., KOZICKI, R., PODBIELSKI, A., THIEMANN, W. (1980) Proc. World Congress Foodborne Infections and Intoxications Berlin (West), p. 469

31. HEIKES, D. L., HOPPER, M. L. (1986) J. Assoc. Off. Anal. Chem. **69**: 990

32. ETIEVANT, P., MAARSE, H., VAN DEN BERG, F. (1986) Chromatographia, **21**: 379–386

33. KOLB, B., AUER, M., POSPISIL, P. (1983) Proc. 5th Int. Symp. on Capillary Chromatogr., Elsevier, Amsterdam, p. 441

34. KOLB, B. (1982) Chromatographia **15**: 587

35. KOLB, B., POSPISIL, P., AUER, M. (1984) Chromatographia **19**: 113

36. DENNISON, J. L., BREDER, C. V., MCNEAL, T., SNYDER, R. C., ROACH, J. A. SPHON, J. A. (1978) J. Assoc. Off. Anal. Chem. **61**: 813

37. UHLER, A. D., MILLER, L. J. (1988) J. Agric. Food Chem. **36**: 772

38. ENTZ, R. C., HOLLIFIELD, H. C. (1982) J. Agric. Food Chem. **30**: 84

39. DE VRIES, J. W., BROGE, J. M., SCHROEDER, J. P., BOWERS, R. H., LARSON, P. A., BURNS, N. M. (1985) J. Assoc. Off. Anal. Chem. **68**: 1112

40. KING, J. R., BENSCHOTER, C. A., BURDITT, A. K. Jr., (1981) J. Agric. Food Chem. **29**: 1003

41. PRANOTO-SOETARDHI, L. A., RIJK, M. A. H., DE KRUIJF, N., DE VOS, R. H. (1986) Int. J. Environ. Anal. Chem. **25**: 151

42. NEUMANN, L., WAGNER, J. (1974) Monatsschr. f. Brauerei **27**: 137

43. HEIKES, D. L. (1987) J. Assoc. Off. Anal. Chem. **70**: 176

44. VIETHS, S., BLAAS, W., FISCHER, M., KRAUSE, C., MATISSEK, R., MEHLITZ, I., WEBER, R. (1988) Z. Lebensm. Unters. Forsch. **186**: 393

45. VIETHS, S., BLAAS, W., FISCHER, M., KLEE, T., KRAUSE, C., MATISSEK, R. ULLRICH, D., WEBER, R. (1988), Dtsch Lebensm. Rundsch. **84**: 381

46. SCUDAMORE, K. A., HEUSER, S. G. (1973) Pest. Sci. **4**: 1

47. GILBERT, J., STARTIN, J. R., CREWS, C. (1985) Food Additives and Contaminants **2**: 55

48. "Methods for the determination of Organic Compounds in finished Drinking Water and Raw Source Water", Physical and Chemical Methods Branch, Environmental Montoring and Support Laboratory, U. S. Environmental Protection Agency, Cincinnati, Ohio 45268 (Revised Nov. 1985)

49. BAUER, U. (1981) Zbl. Bakt. Hyg. I Abt. Orig. B. **174**: 15

50. BAUER, U. (1981) Zbl. Bakt. Hyg. I Abt. Orig. B. **174**: 39

51. BAUER, U. (1981) Zbl. Bakt. Hyg. I Abt. Orig. B. **174**: 200

52. BAUER, U. (1981) Zbl. Bakt. Hyg. I Abt. Orig. B. **174**: 556

53. BAUER, U., SELENKA, F. (1982) Vom Wasser 59: 7, Verlag Chemie, Weinheim/Bergstr.

54. LAHL, U., CETINKAYA, M., V. DÜSZELN, J., STACHEL, B., THIEMANN, W. (1981) Sci. Tot. Environ. **20**: 171

55. HEIKES, D. L. (1985) J. Assoc. Off. Anal. Chem. **68**: 431

56. HEIKES, D. L. (1985) J. Assoc. Off. Anal. Chem. **68**: 1108

57. HEIKES, D. L. (1987) J. Assoc. Off. Anal. Chem. **70**: 215

58. EASLEY, D. M., KLOEPFER, R. D., CARAŠEA, A. M. (1981). J. Assoc. Off. Anal. Chem. **64**: 653

59. NOSTI VEGA, M., GUTIERREZ ROSALES, F., GUTIERREZ GONZALES-QUIJANO, R. (1970) Grasas y Aceites **21**: 276

60. BELLUCO L., PALLAZZO, A., CONTON, C., TONINELLI, G. (1979) Rivista Italiana delle Sostanze Grasse **56**: 91

61. MACHATA, G. (1964) Microchim. Acta **262**: 262

62. MACHATA, G. (1967) Blutalkohol **4**: 3

63. ROTTSAHL, H., JESSEN, T. (1987) Deutsche Lebensmittel-Rundschau **83**: 42

64. HOLLINGWORTH, JR. T. A., THROM, H. R. (1982) J. Food Sci. **47**: 1315

65. GRAMCCIONI, L. MILANA, M. R., DI MARZIO, S., LORUSSO, S. (1986) Chromatographia **21**: 9

66. VAN HAVER, W., GORDTS, L. (1977) Z. Lebensm. Unters. Forsch. **165**: 28

67. BLAICHER, G., WOIDICH, H., PFANNHAUSER, W. (1980) Ernährung **4**: 440

68. U. K. Ministry of Agriculture, Fisheries and Food, Committee for Analytical Methods for Residues of Pesticides and Veterinary Products in Foodstuffs (1982) Analyst **106**: 782

69. EISENBRAND, G. (1981) "N-Nitrosoverbindungen in Nahrung und Umwelt; Eigenschaften, Bildungswege, Nachweisverfahren und Vorkommen", Wissenschaftliche Verlagsgesellschaft, Stuttgart

70. BILLEDEAU, S. M. THOMSON, JR., H. C., HANSEN, E. B., MILLER, B. J. (1984) J. Assoc. Off. Anal. Chem. **67**: 557

71. DECADT, G., BAEYENS, W., BRADLEY, D. GOEYENS, L. (1985) Analytical Chemistry **57**: 2788

72. KOLB, B., POSPISIL, P., AUER, M. (1981) J. Chromatogr. **204**: 371

73. ZENNER, H. F., KRZEMINSKI, C. (1984) Zeitschr. Ges. Hygiene Grenzgeb. **30**: 622

74. GROTE, H., LEUGERS, G. (1987) Fresenius Z. Anal. Chem. **327**: 782

75. BILTCLIFFE, D. O., WOOD, R. (1982) J. Assoc. Publ. Anal. **20**: 55

76. KROCKENBERGER, H., GMERCK, D. (1987) Fresenius Z. Anal. Chem. **327**: 55

77. CHUDY, J. LC., CROSBY, N. T. (1977) Food Cosmet. Toxicol. **15**: 547

78. DI PASQUALE, G., DI IORIO, G., CAPACCIOLI, T., GAGLIARDI, P., VERGA, G. R. (1978) J. Chromatogr. **160**: 133

79. HOLLIFIELD, R. C., BREDER, C. V. DENNISON, J. L., ROACH, J. A. G., ADAMS, W. S. (1980) J. Assoc. Off. Anal. Chem. **63**: 173

80. STARTIN, J. R., GILBERT, J. (1984) J. Chromatogr. **294**: 427

81. MCNEAL, T. P., BREDER, C. V. (1987) J. Assoc. Off. Anal. Chem. **70**: 18

82. SANTA MARIA, I., CARMI, J. D., OBER, A. G. (1986) Bull. Environ. Contam. Toxicol. **37**: 207

83. VARNER, S. L., BREDER, C. V. (1984) J. Assoc. Off. Anal. Chem. **67**: 516

84. RICOTTILLI, F., BRANCA, P., MARTIRE, N. (1981) Bollettino dei Chimici dei Laboratori Provinciali **32**: 133

85. DAHLGRAN, J. R., SHINGLETON, C. R. (1987) J. Assoc. Off. Anal. Chem. **70**: 796

Index

References

Alles, G.; Bauer, U.; Selenka, F. (1981): Zentralblatt Bakteriologische Hygiene I, Abt. Orig. B 174:241; Gustav Fischer Verlag, Stuttgart, Fig. III.3.-18, 305 — Atlas of Chromatograms, GC 59, in Ettre, L. S. (Atlas Consultant) (1988): J. Chromatographic Science 26:190; Preston Publications, A Division of Preston Industries, Inc., Niles USA, Fig. II.5.-9, 190 — Badings, H. T.; de Jong, C.; Dooper, R. P. M. (1985): J. High Res. Chromatogr. 8:755; Alfred Hüthig Verlag GmbH, Heidelberg, Fig. III.3.-3, 283 — Beck, H.; Eckhart, K.; Mathar, W.; Wittkowski, R. (1986): Chemosphere 18:417–424; Pergamon Press, Oxford, Fig. III.2.-17, 268 — Bio-Rad: The Analysis of Triglycerides on RSL-300, Technisches Informationsblatt 701; Firmenschrift, Nazareth, Belgium, Fig. II.5.-4, 184 — Blaicher, G.; Woidich, H.; Pfannhauser, W. (1980): Ernährung 4:440; Fachzeitschriftenverlagsgesellschaft, Wien, Fig. III.3.-21, 310 — Bomberg, S.: Roeraade, J. (1989): Food Chemistry; Elsevier Publisher, Ltd., Barking, Fig. II.3.-17, 136 — Bretschneider, W.; Werkhoff, P. (1988): J. High Res. Chromatogr. 11:543; Alfred Hüthig Verlag GmbH, Heidelberg, Fig. II.3.-4, 114; Fig. II.3.-5, 115 — Chadha, R. K.; Lawrence, J. F.; Conacher, H. B. S. (1986); Journal of Chromatography 356:444; Elsevier Science Publishers B. V., Amsterdam, Fig. II.2.-10, 105 — Chrompack, International B. V. (1987); Chrompack News 14 (1):5; Firmenschrift, Middelburg, Fig. II.5.-10, 191 — Chrompack, International B. V. (1987): Chrompack Packard Triglyceride Analyzer, Firmenschrift; Middelburg, Fig. II.5.-3, 183 — Chrompack, International B. V. (1988): Chrompack News 15 (2):5; Firmenschrift, Middelburg, Fig. II.5.-6, 186 — Clark, B. J., Jr.; Chamblee, T. S.; Jacobucci, G. A. (1987): J. Agric. Food Chem. 35:514; American Chemical Society, Washington, Fig. II.3.-13, 131 — Cobelli, L.; Banfi, S. (1987): International Laboratory 9:46; International Scientific Communication Corporate, Shelton, Fig. III.3.-5, 284 — Decadt, G.; Baeyens, W.; Bradley, D.; Goeyens, L. (1985): Analytical Chemistry 57, 14:2789; Amercian Chemical Society, Washington, Fig. III.3-25, 314 — Deleu, R.; Copin, A. (1984): J. High Res. Chromatogr. 7:338; Alfred Hüthig Verlag GmbH, Heidelberg, Fig. III.1.-2, 216 — Entz, R. C.; Hollifield, H. C. (1982): J. Agric. Food Chem. 30:84; American Chemical Society, Washington, Fig. III.3.-17, 303 — Firestone, D.; Horowitz, W. (1984): Journal of Chromatography 309:254; Elsevier Science Publishers B. V., Amsterdam, Fig. II.2.-4, 95 — Fischer Labor- und Verfahrenstechnik (1987): Curie-Punkt-Pyrolyse und automatischer Probengeber für GC-, IR- oder MS-Anwendung p. 3, Fig. 2; Firmenschrift, Meckenheim, Fig. II.5.-2, 180 — Flath, R. A.; Forrey, R. R. (1970): J. Agric. Food Chem. 18:306, American Chemical Society, Washington, Fig. II.3.-8, 117 — Fürst, P.; Krüger, C.; Meemken, H.-A.; Groebel, W. (1987): Journal of Chromatography 405:311–317; Elsevier Science Publishers B. V., Amsterdam, Fig. III.2.-7, 250 — Fürst, P.; Krüger, C.; Meemken, H.-A.; Groebel, W. (1987): Z. Lebensm. Unters. Forsch. 185:394–397; Springer Verlag GmbH & Co. KG, Berlin, Fig. III.2.-9, 252 — Greenberg, M. J.; Hoholick, J.; Robinson, R.; Kubis, K.; Groce, J.; Weber, L. (1984): Journal of Food Science 49 (6):1623; Institute of Food

Technologists, Chicago, Fig. II.1.-3, 77 — Grob, K.; Grob, K., Jr. (1978): Journal of Chromatography 151:311; Elsevier Science Publishers B. V., Amsterdam, Fig. II.3.-2, 111; Fig. II.3.-3, 112 — Grob, K.; Grob, G. (1979): High Res. Chromatogr. 2:109; Alfred Hüthig Verlag GmbH, Heidelberg, Fig. II.3.-1, 110 — Grob, K.; Grob, G. (1982): J. High Res. Chromatogr. 5:349; Alfred Hüthig Verlag GmbH, Heidelberg, Fig. II.3.-12, 126 — Hardt, R.; Baltes, W. (1987): Z. Lebensm. Unters. Forsch. 185:278; Springer Verlag GmbH & Co. KG, Berlin, Fig. II.5.-14, 198-199 — Hardt, R.; Baltes W. (1987): Z. Lebensm. Unters. Forsch. 185:279; Springer Verlag GmbH & Co. KG, Berlin, Fig. II.5.-15, 200 — Hinshaw, J. V.; Ettre, L. S. (1988): Proc. Ninth Int. Symp. Capillary Chromatography p. 609; Alfred Hüthig Verlag GmbH, Heidelberg, Fig. II.5.-1, 180 — ict- Handelsgesellschaft m. b. H. (1987): J & W Scientific General Catalog 1987/88, p. 29; Frankfurt, Fig. II.5.-8, 189 — Kolb, B. (1986): Gaschromatographie mit Kapillartrennsäulen (Günther, W.; Schlegelmilch, W. F. [Eds.]), 149; Vogel Verlag, Würzburg, Fig. III.3.-4, 283 — Kolb, B.; Liebhardt, B.; Ettre, L.S. (1986): Chromatographia 21, 6:305, p. 306; Friedrich Vieweg & Sohn Verlagsgesellschaft, Wiesbaden, Fig. III.3.-2, 281 — Littmann, S.; Acker, L.; Schulte, E. (1982): Z. Lebensm. Unters. Forsch., 175:109; Springer Verlag GmbH & Co. KG, Berlin, Fig. II.2.-6, 99 — Low, N. H.; Sporns, P. (1988): Journal of Food Science, 53 (2):560; Institute of Food Technologists, Chicago, Fig. II.5.-11, 193; Fig. II.5.-12, 194 — Manninen, A.; Kuitunen, M. L.; Julin, L. (1987): Journal of Chromatography 394:465; Elsevier Science Publishers B. V., Amsterdam, Fig. III.1.-6, 222 — Meemken, H.-A.; Fürst, P.; Habersaat, K. (1982): Deutsche Lebensmittelrundschau 78:282-287; Wissenschaftliche Verlagsgesellschaft, Stuttgart, Fig. III.1.-15, 234 — Meemken, H.-A.; Rudolph, P.; Fürst, P.; (1987): Deutsche Lebensmittelrundschau 83:239-245; Wissenschaftliche Verlagsgesellschaft, Stuttgart, Fig. III.1.-12, 230 — Nitz, S.; Drawert, F.; Julich, E. (1984): Chromatographia 18:313; Friedrich Vieweg & Sohn Verlagsgesellschaft, Wiesbaden, Fig. II.3.-7, 116 — Poy, F.; Visani, S.; Terrosi, F. (1981): Journal of Chromatography 217:81; Elsevier Science Publishers B. V., Amsterdam, Fig. II.3.-6, 116 — Preuss, A.; Thier, H.-P. (1982): Z. Lebensm. Unters. Forsch. 175:97; Springer Verlag GmbH & Co. KG, Berlin, Fig. II.1.-5, 80 — Preuss, A.; Thier, H.-P. (1982): Z. Lebensm. Unters. Forsch. 175:96; Springer Verlag GmbH & Co. KG, Berlin, Fig. II.5.-13, 195 — Schulte, E. (1983): Praxis der Kapillar-Gaschromatographie 91; Springer Verlag GmbH & Co. KG, Berlin, Fig. II.5.-5, 185 — Stan, H.-J.; Mrowetz, D. (1983): J. High Res. Chromatogr. 6:255; Alfred Hüthig Verlag GmbH, Heidelberg, Fig. III.1.-7, 223 — Takeoka, G. R.; Flath, R. A.; Guentert, M.; Jennings, W. (1988): J. Agric. Food Chem. 36:553; American Chemical Society, Washington, Fig. II.3.-10, 122 — Takeoka, G. R.; Guentert, M.; Macku, C.; Jennings, W. (1986): Biogeneration of Aromas (Parliment, T. H.; Croteau, R., [Eds.]), ACS Symposium Series 317, American Chemical Society, Washington, Fig. II.3.-14, 133 — Thier, H.-P.; Frehse, H. (1986): Rückstandsanalytik von Pflanzenschutzmitteln; Georg Thieme Verlag, Stuttgart, Fig. III.1.-1, 208 — Toulemonde, B.; Richard, H. M. J. (1983): J. Agric. Food Chem. 31:365; American Chemical Society, Washington, Fig. II. 3.-11, 124 — Veith, H. J.; Fischer, M.; Hua, J.; Roth, H. (1986): Deutsche Lebensmittelrundschau 82 (8): 257;

Wissenschaftliche Verlagsgesellschaft, Stuttgart, Fig. II.1.–9, 87 — Vieths, S.; Blaas, W.; Fischer, M.; Klee, T.; Krause, L.; Matissek, R.; Ullrich, D.; Weber, R. (1988): Deutsche Lebensmittelrundschau 84:381–388; Wissenschaftliche Verlagsgesellschaft, Stuttgart, Fig. III.3.–14, 299 — Wittkowski, R.; Baltes, W.; Takeoka, G.; Jennings, W. (1988): J. High Res. Chromatogr. 2:109; Alfred Hüthig Verlag GmbH, Heidelberg, Fig. II.3.–16, 134 — Yamaguchi, K.; Shibamoto, T. (1981): J. Agric. Food Chem. 29:366; American Chemical Society, Washington, Fig. II.3.–9, 120 —